The Organic Waste Revolution: Composting Solutions for Every Household

Jessica

Copyright © [2023]

Title: The Organic Waste Revolution: Composting Solutions for Every Household
Author's: Jessica

This book was printed and published by [Publisher's: **Jessica**] in [2023]

ISBN:

TABLE OF CONTENT

Chapter 4: Composting Techniques for Every Household

Backyard Composting

Building a Basic Compost Pile

Utilizing Compost Bins and Tumblers

Indoor Composting

Vermicomposting with Worms

Bokashi Composting

Community Composting

Participating in Local Composting Programs

Establishing Community Composting Initiatives

Chapter 5: Troubleshooting Common Composting Issues

Dealing with Odor Problems

Managing Pests and Rodents

Addressing Excessive Moisture or Dryness

Troubleshooting Slow or Incomplete Composting

Chapter 6: Utilizing Compost in Your Household   45

Chapter 7: Taking Organic Waste Recycling Further   56

Chapter 1: Introduction to Organic Waste Recycling

The Importance of Recycling Organic Waste

In today's world, where environmental concerns are at the forefront of global discussions, it is crucial for individuals from all walks of life to embrace sustainable practices. One such practice that holds immense significance is the recycling of organic waste. This subchapter will delve into the importance of recycling organic waste and how it can positively impact our environment.

Organic waste refers to any material that originates from plants or animals and is biodegradable. This waste includes food scraps, yard trimmings, paper products, and even animal manure. When these materials end up in landfills, they undergo anaerobic decomposition, releasing harmful greenhouse gases such as methane. By recycling organic waste, we can divert it from landfills and harness its potential to create nutrient-rich compost.

Recycling organic waste has numerous benefits, both on an individual and global scale. First and foremost, it significantly reduces the amount of waste going to landfills. According to studies, organic waste constitutes a significant portion of municipal solid waste, contributing to the ever-growing landfill crisis. By recycling organic waste, we can alleviate the strain on landfills and extend their lifespan.

Furthermore, recycling organic waste through composting offers an effective solution to soil degradation and erosion. Compost is a natural fertilizer that enriches the soil with essential nutrients, improving its structure and water-holding capacity. By adding compost to gardens, farms, and green spaces, we can enhance plant growth, reduce the need for chemical fertilizers, and mitigate soil erosion.

Recycling organic waste also plays a vital role in mitigating climate change. As mentioned earlier, when organic waste decomposes in landfills, it releases methane, a potent greenhouse gas. By diverting this waste from landfills and composting it instead, we can significantly reduce methane emissions. This action directly contributes to mitigating climate change and creating a more sustainable future for generations to come.

Moreover, recycling organic waste fosters a sense of responsibility and environmental awareness among individuals. Engaging in composting not only reduces waste but also encourages people to be mindful of their consumption habits. It instills a sense of connection with nature and promotes sustainable living practices, which are essential for a healthier planet.

In conclusion, recycling organic waste is of paramount importance for environmental sustainability. By diverting organic waste from landfills and composting it instead, we can reduce greenhouse gas emissions, mitigate soil erosion, and promote sustainable living practices. It is a simple yet effective solution that every individual can embrace, regardless of their background or lifestyle. Let us all join the organic waste revolution and contribute to a greener, healthier future for ourselves and generations to come.

Benefits of Composting for the Environment

Composting is a natural process in which organic waste materials, such as food scraps, yard trimmings, and other biodegradable substances, are broken down and transformed into nutrient-rich soil. In recent years, the importance of recycling and finding sustainable solutions to manage organic waste has become increasingly evident. This subchapter will explore the benefits of composting for the environment and highlight the role it plays in the broader context of recycling.

First and foremost, composting helps reduce the amount of waste that ends up in landfills. Organic waste, when disposed of in landfills, decomposes anaerobically, releasing harmful greenhouse gases such as methane. Methane is a potent contributor to climate change, and by diverting organic waste from landfills, composting significantly reduces these emissions. It acts as a natural carbon sink, trapping carbon dioxide and preventing its release into the atmosphere.

Moreover, composting enriches soil health and fertility. The resulting compost, often referred to as "black gold," is a nutrient-rich substance that enhances soil structure, moisture retention, and overall plant growth. By using compost in gardens, farms, and landscaping projects, we can reduce the reliance on synthetic fertilizers, which can have detrimental effects on the environment and human health. Compost acts as a natural alternative, providing plants with essential nutrients while promoting biodiversity and reducing the need for harmful chemicals.

Another crucial benefit of composting is its ability to conserve water. Compost improves soil's water-holding capacity, reducing the need for excessive irrigation. In a world where water scarcity is becoming a pressing issue, composting offers a practical solution by improving soil's water retention abilities and reducing water waste.

Additionally, composting fosters a sense of environmental stewardship and community engagement. By participating in composting initiatives, individuals actively contribute to reducing their carbon footprint and promoting sustainable practices. Composting can be done at various scales, from small household bins to community-wide programs, allowing everyone to play a role in the organic waste revolution.

In conclusion, composting offers numerous benefits for the environment. It helps reduce greenhouse gas emissions, enriches soil health, conserves water, and promotes sustainable practices. By embracing composting as a recycling solution for organic waste, we can make a significant positive impact on our planet and create a more sustainable future for generations to come.

Understanding the Process of Composting

Composting is a simple yet effective method of recycling organic waste that has gained significant popularity in recent years. It offers an environmentally friendly solution for reducing waste and producing nutrient-rich soil amendments. In this subchapter, we will explore the process of composting and its significance in the larger context of recycling.

Composting involves the decomposition of organic materials such as kitchen scraps, yard waste, and even paper products. Through a natural process called aerobic decomposition, microorganisms break down these materials, resulting in a nutrient-rich substance known as compost. This compost can then be used to enrich the soil in gardens, farms, or even potted plants.

The process of composting can be broken down into several stages. It begins with the collection of organic waste, which can include fruit and vegetable peels, coffee grounds, eggshells, leaves, grass clippings, and more. Once collected, these materials are mixed together in a compost bin or pile, providing the ideal conditions for decomposition to occur.

The next crucial step is providing the right balance of carbon and nitrogen-rich materials. Carbon-rich materials, often referred to as "browns," include dry leaves, wood chips, and shredded paper. Nitrogen-rich materials, known as "greens," include fresh grass clippings, vegetable scraps, and coffee grounds. Achieving the right balance between browns and greens helps create an optimal environment for microorganisms to thrive and break down the organic matter.

Composting also requires the right amount of moisture and oxygen. Microorganisms responsible for the decomposition process require both water and oxygen to function effectively. It is essential to maintain proper moisture levels by periodically watering the compost pile and turning it to ensure adequate aeration.

The importance of composting in the larger context of recycling cannot be overstated. Organic waste, when sent to landfills, produces harmful greenhouse gases such as methane, contributing to climate change. By composting organic waste, we can divert a significant portion of this waste away from landfills, reducing greenhouse gas emissions and helping combat climate change.

Additionally, composting enriches the soil with vital nutrients, improves soil structure, and enhances water retention capacity. This, in turn, promotes healthier plant growth, reduces the need for chemical fertilizers, and helps conserve water.

In conclusion, understanding the process of composting is crucial for every individual interested in the importance of recycling. By practicing composting in our households, we can contribute to reducing waste, mitigating climate change, and creating healthier and more sustainable environments. Whether you have a small balcony or a large backyard, composting is an accessible and rewarding way to make a positive impact on our planet.

Chapter 2: Getting Started with Organic Waste Recycling

Assessing Your Household's Organic Waste

In today's world, where environmental concerns are more prominent than ever, it is crucial for every household to play its part in recycling and reducing waste. One significant aspect of this is properly managing organic waste. Organic waste refers to any waste that comes from plants or animals, and it can be recycled through composting, a natural process that transforms organic materials into nutrient-rich soil. Assessing your household's organic waste is an essential step towards adopting sustainable practices and making a positive impact on the environment.

Why is assessing your organic waste important? Firstly, understanding the amount and types of organic waste generated in your household allows you to identify areas where waste can be reduced or recycled. By doing so, you can contribute to reducing the overall waste going to landfills, which in turn decreases greenhouse gas emissions. Additionally, composting organic waste at home provides you with a valuable resource – nutrient-rich compost that can be used to nourish your garden or potted plants. This reduces the need for synthetic fertilizers, minimizing chemical run-off into water sources and promoting healthier plant growth.

To assess your household's organic waste, start by conducting a waste audit. This involves keeping track of the organic waste you generate over a designated period, such as a week or a month. Categorize the waste into different types, such as fruit and vegetable scraps, coffee grounds, eggshells, and yard waste. This audit will give you a clear picture of the volume and variety of organic waste you produce.

Once you have gathered this information, you can begin implementing effective waste management strategies. Start by setting up a composting system in your backyard or opting for a small-scale indoor composting bin if you lack outdoor space. By diverting organic waste from the landfill and into compost, you are not only reducing waste but also creating a valuable resource for your garden.

Furthermore, consider reducing organic waste by making mindful choices in your shopping habits. Buy only what you need, avoiding excessive food waste. Additionally, explore ways to repurpose certain food scraps, such as using vegetable peels for homemade broths or making compost tea from coffee grounds.

Assessing your household's organic waste is a crucial step in embracing the importance of recycling. By understanding and managing your organic waste, you contribute to a cleaner, greener future for our planet. So, take charge of your household's organic waste and join the organic waste revolution today!

Setting Up a Composting System

Composting is a simple yet powerful way to reduce organic waste and contribute to a greener planet. In this subchapter, we will dive into the process of setting up a composting system right in your own home. Whether you live in a house with a backyard or an apartment with limited space, there are composting solutions suitable for everyone.

The importance of recycling cannot be emphasized enough in today's world. With landfills overflowing and limited resources, it is crucial to find sustainable ways to manage our waste. Composting offers an effective solution by turning organic waste into nutrient-rich soil, which can be used to support plant growth and reduce the need for chemical fertilizers. By implementing a composting system, you not only contribute to the environment but also create a positive impact on your household and community.

The first step in setting up a composting system is determining the suitable method for your space and lifestyle. If you have a backyard, a traditional outdoor compost pile or bin may be the best option. This method allows you to compost a variety of materials such as fruit and vegetable scraps, coffee grounds, yard waste, and even paper products. It is essential to create a balance between brown materials (carbon-rich) and green materials (nitrogen-rich) to ensure proper decomposition.

For those living in apartments or with limited outdoor space, indoor composting systems are a fantastic alternative. These systems, such as worm composting or bokashi composting, can be set up in small containers and don't produce any foul odors. They are ideal for composting kitchen scraps and some paper products. The end result is a nutrient-dense fertilizer that can be used for potted plants or even shared with local community gardens.

Once you have chosen the suitable method, it's time to start composting! Remember to chop larger materials into smaller pieces to speed up the decomposition process. Regularly turn or mix the compost to provide oxygen and promote decomposition. Keep the compost moist but not too wet to avoid unpleasant smells. With time and patience, you will start to see the magic of composting as the waste transforms into dark, crumbly soil.

In conclusion, setting up a composting system is a practical and rewarding way to contribute to the organic waste revolution. By recycling our organic waste, we can reduce landfill waste, conserve resources, and create nutrient-rich soil for our plants. Whether you have a backyard or live in an apartment, there is a composting solution for everyone. Start your composting journey today and become a part of the movement towards a more sustainable future.

Choosing the Right Composting Method for Your Household

In today's world, where the importance of recycling and sustainable living is becoming increasingly evident, composting has emerged as a simple yet powerful solution for managing organic waste. Composting not only helps reduce the amount of waste that ends up in landfills but also enriches the soil, conserves water, and reduces the need for chemical fertilizers. However, with various composting methods available, it can be overwhelming to determine which one is best suited for your household. This subchapter aims to guide everyone in choosing the right composting method that aligns with their unique needs and lifestyle.

There are several factors to consider when selecting a composting method. Firstly, the available space in your household plays a crucial role. If you have limited outdoor space, you may opt for indoor or small-scale composting methods such as vermicomposting or bokashi composting. These methods utilize worms or microbes to break down organic waste, making them perfect for apartment dwellers or those with minimal yard space.

Furthermore, the types of organic waste you generate should be considered. If you primarily have kitchen scraps and yard trimmings, traditional backyard composting using a compost bin or pile might be the best fit. This method involves layering green and brown materials, allowing them to decompose naturally over time. On the other hand, if you have an abundance of food scraps and want a faster composting process, consider investing in a tumbling composter or an electric compost bin that accelerates decomposition.

Additionally, your level of involvement and commitment to composting should be taken into account. If you enjoy getting your hands dirty and actively participating in the composting process, a manual method like hot composting might be ideal. This method requires regular turning and monitoring of the compost pile to ensure proper aeration and decomposition. Conversely, if you prefer a more hands-off approach, a self-contained composting system that requires minimal maintenance, such as an aerobic compost bin, may be the right choice.

By carefully evaluating your available space, the type of organic waste you generate, and your level of involvement, you can choose a composting method that seamlessly integrates into your household routine. Remember, the right composting method is one that you can sustain over the long term. Embrace the organic waste revolution and make a positive impact on the environment by selecting the composting method that suits you best. Together, we can contribute to a greener and more sustainable future for all.

Chapter 3: Composting Basics

Essential Components for Successful Composting

Composting is a simple and effective way to reduce organic waste and create nutrient-rich soil for your plants. Whether you are a seasoned gardener or someone looking to make a positive impact on the environment, understanding the essential components for successful composting is crucial. In this subchapter, we will explore the key elements that contribute to a thriving composting system, helping you harness the power of organic waste and contribute to the importance of recycling.

1. Organic Waste: The foundation of successful composting lies in the organic waste you collect. This can include kitchen scraps like fruit and vegetable peels, coffee grounds, and eggshells, as well as yard waste such as leaves, grass clippings, and small branches. By diverting this waste from landfills, you are not only reducing methane emissions but also enriching your compost pile.

2. Carbon and Nitrogen Ratio: Achieving the right balance of carbon-rich "browns" and nitrogen-rich "greens" is vital for a healthy compost. Browns include materials like dried leaves, straw, and shredded newspaper, while greens consist of fresh grass clippings, vegetable scraps, and manure. Aim for a ratio of roughly 3 parts browns to 1 part greens to ensure proper decomposition and prevent odors.

3. Moisture: Adequate moisture is essential for composting. Your compost pile should be moist but not overly wet. Regularly water the pile, especially during dry periods, to maintain a moisture level similar

to a damp sponge. This ensures that the microorganisms responsible for breaking down the organic matter can thrive and work efficiently.

4. Oxygen: Composting is an aerobic process, meaning it requires oxygen. To ensure proper airflow, periodically turn or aerate your compost pile. This helps to prevent the formation of compacted layers and allows oxygen to reach the microorganisms, facilitating decomposition.

5. Temperature: Proper temperature control is crucial for successful composting. As organic matter decomposes, heat is generated. Aim for a temperature range between 110°F (43°C) and 160°F (71°C) to optimize decomposition and kill any potential pathogens or weed seeds.

By understanding and implementing these essential components, you can create a thriving composting system in your own household. This not only reduces your impact on landfills but also provides you with nutrient-rich compost that will nourish your plants and gardens. Composting is an integral part of the broader importance of recycling, as it closes the loop on organic waste and helps create sustainable, environmentally friendly practices. Whether you have a small backyard or live in an apartment, everyone can make a significant contribution to the organic waste revolution.

Balancing Carbon and Nitrogen in Your Compost

Composting is an essential part of the organic waste revolution, offering a simple and effective way to recycle our kitchen scraps and yard waste. By composting, we can reduce the amount of waste that ends up in landfills and create nutrient-rich soil to nourish our gardens. However, for successful and efficient composting, it is crucial to understand the balance between carbon and nitrogen in your compost pile.

Carbon and nitrogen are the two main components required for composting. Carbon-rich materials provide energy for the microorganisms that break down the organic matter, while nitrogen-rich materials supply the essential proteins needed for their growth and reproduction. Achieving the right balance between these two elements is crucial for the decomposition process and the overall quality of your compost.

Carbon-rich materials, often referred to as "browns," include items such as dried leaves, straw, wood chips, and cardboard. These materials are high in carbon and provide the energy needed for the microorganisms to thrive. Nitrogen-rich materials, or "greens," consist of fresh grass clippings, fruit and vegetable scraps, coffee grounds, and even manure. These materials are high in nitrogen and provide the necessary proteins for microbial growth.

To achieve a balanced compost pile, aim for a ratio of approximately 30 parts carbon to 1 part nitrogen. This balance ensures that the microorganisms have enough energy and nutrients to break down the organic matter effectively. Too much carbon will result in a slow and inefficient decomposition process, while an excess of nitrogen can lead to a smelly and slimy compost pile.

It's important to note that the carbon-to-nitrogen ratio can vary depending on the specific materials you use. For example, woody materials such as sawdust or wood chips have a higher carbon content, so you might need to add more nitrogen-rich materials to maintain the balance. On the other hand, fresh grass clippings have a high nitrogen content, so adding more carbon-rich materials like dried leaves can help balance the pile.

Regularly monitoring and adjusting the carbon-to-nitrogen ratio in your compost pile is essential for successful composting. Turning the pile regularly to ensure proper aeration and mixing the materials can also help maintain an optimal balance. By achieving the right carbon-to-nitrogen ratio, you can create compost that is rich in nutrients, free from foul odors, and perfect for nourishing your plants and gardens.

In conclusion, balancing carbon and nitrogen in your compost is essential for successful and efficient composting. By understanding the carbon-to-nitrogen ratio and using a mix of carbon-rich and nitrogen-rich materials, you can create high-quality compost that will enrich your soil and reduce waste. Join the organic waste revolution by embracing composting and making a positive impact on the environment and your own garden.

Maintaining Proper Moisture Levels

One crucial aspect of successful composting is maintaining proper moisture levels in your compost pile. Moisture plays a vital role in ensuring the organic waste breaks down efficiently and becomes nutrient-rich compost. In this subchapter, we will explore the significance of maintaining optimal moisture levels and provide practical tips to help you achieve the best results in your composting journey.

Why is Moisture Important in Composting?

Moisture is essential because it facilitates the decomposition process by providing a suitable environment for microorganisms to thrive. The right amount of moisture ensures that bacteria, fungi, and other decomposers can break down the organic matter effectively. Without adequate moisture, these microorganisms cannot survive and the composting process slows down or stops altogether.

When the compost pile lacks moisture, it becomes dry and oxygen levels decrease, leading to the growth of anaerobic microorganisms. These anaerobic conditions produce an unpleasant smell and result in the production of harmful gases, such as methane. Maintaining proper moisture levels in your compost pile not only ensures a productive composting process but also prevents odors and reduces greenhouse gas emissions.

Tips for Maintaining Proper Moisture Levels

To maintain optimal moisture levels in your compost pile, follow these helpful tips:

1. Monitor the moisture: Regularly check the moisture content of your compost pile by squeezing a handful of compost. It should feel like a damp sponge, releasing a few drops of water when squeezed. If it feels too dry, add water, and if it's too wet, add dry organic materials like leaves or shredded paper to absorb the excess moisture.

2. Provide proper drainage: Ensure that your composting system has good drainage to prevent waterlogging. A compost bin or pile should have holes or spaces at the bottom to allow excess moisture to drain out.

3. Cover the pile: Use a tarp or cover your compost pile to protect it from heavy rainfall, which can make the pile excessively wet. A cover also helps to retain moisture during dry periods.

4. Turn the pile: Regularly turning your compost pile helps distribute moisture evenly throughout the materials. This prevents dry pockets from forming and encourages decomposition.

5. Use kitchen scraps wisely: Kitchen scraps like fruit and vegetable peels contain high moisture content. While they are excellent compost ingredients, be mindful of the moisture they contribute and balance them with drier materials like cardboard or straw.

By maintaining proper moisture levels in your compost pile, you'll create an optimal environment for decomposition, ensuring the production of nutrient-rich compost. Remember, composting is a simple and effective way to recycle organic waste, reducing landfill waste and contributing to a healthier planet.

Managing Temperature and Airflow

One of the key aspects of successful composting is managing the temperature and airflow within your composting system. This subchapter aims to provide you with essential information on how to maintain the optimum conditions for organic waste decomposition, ensuring a successful composting process.

Temperature plays a crucial role in the breakdown of organic matter. A well-regulated compost pile should ideally maintain a temperature between 120°F and 160°F (49°C to 71°C). This range ensures efficient decomposition and kills off harmful pathogens and weed seeds. To achieve and maintain this temperature, proper airflow is essential.

Airflow is necessary to ensure oxygen availability for the microorganisms responsible for decomposition. Without adequate oxygen, the composting process can become anaerobic, leading to unpleasant odors and slow decomposition. To promote airflow, it is recommended to turn or mix your compost pile regularly. This can be done using a pitchfork or a compost turning tool. By aerating the pile, you allow fresh oxygen to reach the microorganisms, stimulating their activity and maintaining the desired temperature.

Additionally, managing airflow can be achieved by layering your compost pile properly. Alternate between green and brown materials, ensuring a good balance. Green materials such as kitchen scraps and fresh grass clippings provide nitrogen, while brown materials like dried leaves and straw offer carbon. This layering technique facilitates air circulation and prevents the pile from becoming too compacted, which can hinder decomposition.

Monitoring the temperature and airflow in your composting system is crucial. You can use a compost thermometer to check if the temperature is within the desired range. If it falls below, turning the pile and adding more green materials can help raise the temperature. On the other hand, if the temperature is too high, you can add more brown materials or increase the airflow by turning the pile more frequently.

By managing temperature and airflow effectively, you can ensure that your composting process is efficient, odor-free, and successful. Remember, composting is an essential part of the recycling process, as it diverts organic waste from landfills and reduces greenhouse gas emissions. By composting at home, you are making a positive impact on the environment and contributing to a more sustainable future.

In conclusion, understanding the importance of recycling organic waste and the role of temperature and airflow in composting is crucial for every household. By implementing the techniques discussed in this subchapter, you can effectively manage your compost pile and contribute to the organic waste revolution. Start composting today and become a part of the solution!

Chapter 4: Composting Techniques for Every Household

Backyard Composting

In this subchapter, we will explore the fascinating world of backyard composting, a simple and effective way to turn organic waste into nutrient-rich soil for your garden. Composting is not only beneficial for the environment but also for every household. By understanding the importance of recycling and taking small steps towards reducing our waste, we can contribute to a more sustainable future.

Composting is the process of decomposing organic materials, such as kitchen scraps, yard waste, and even certain paper products, into a dark, crumbly substance called compost. This compost can then be used as a natural fertilizer for your plants, reducing the need for chemical-based alternatives. The best part? You can easily do this in your own backyard.

One of the primary reasons why backyard composting is important is because it diverts organic waste from landfills. Organic waste, when buried in landfills, produces methane, a potent greenhouse gas that contributes to climate change. By composting at home, you are reducing the amount of waste that ends up in landfills and minimizing your carbon footprint.

Not only does composting help the environment, but it also benefits your garden. Compost improves soil structure, retains moisture, and provides essential nutrients for plants. It acts as a natural fertilizer, promoting healthy growth and increasing the resistance of plants to pests and diseases. In essence, backyard composting allows you to close the loop and create a sustainable cycle of waste reduction and resource regeneration.

Getting started with backyard composting is easier than you might think. All you need is a compost bin or pile, organic waste, and a little bit of patience. You can compost a wide range of materials, including fruit and vegetable scraps, coffee grounds, tea leaves, eggshells, yard trimmings, and shredded paper. By following a few simple guidelines – such as avoiding meat, dairy, and oily foods – you can ensure that your composting process runs smoothly and effectively.

In conclusion, backyard composting is a powerful tool that every household can utilize to contribute to the organic waste revolution. By understanding the importance of recycling and taking small steps towards reducing our waste, we can make a significant impact on the environment. So, roll up your sleeves, grab a compost bin, and embark on a journey towards sustainable living. Together, we can create a greener, cleaner, and healthier world for ourselves and future generations.

Building a Basic Compost Pile

Composting is a simple and effective way to reduce organic waste and create nutrient-rich soil for your garden. In this subchapter, we will explore the process of building a basic compost pile, suitable for every household. Whether you are a seasoned gardener or a beginner, composting is a key step in the organic waste revolution, and it is essential for everyone to understand its importance in recycling.

To initiate the composting process, start by selecting a suitable location for your compost pile. Find an area in your backyard that is easily accessible and receives a good amount of sunlight. It is important to choose a spot that is away from direct pathways and preferably close to a water source. Once you have identified the perfect spot, gather the necessary materials to build your compost pile.

The key ingredients for a compost pile are known as "greens" and "browns." Greens include fresh grass clippings, vegetable scraps, and coffee grounds, while browns consist of dry leaves, straw, and newspaper. Aim for a ratio of approximately 3 parts browns to 1 part greens. Layer these materials in your compost pile, starting with a thick layer of browns at the bottom, followed by a layer of greens. Repeat this process until you have a pile that is at least 3 feet wide and 3 feet high.

To accelerate the composting process, it is important to provide the pile with sufficient moisture. Use a garden hose to lightly moisten the materials, ensuring they are damp but not waterlogged. Additionally, turning the compost pile every few weeks with a pitchfork or shovel will help aerate the materials and speed up decomposition.

Remember, composting is a natural process, and it may take anywhere from a few months to a year for your compost pile to fully decompose. Regularly monitor the moisture level and temperature of the pile. If it becomes too dry, add water, and if it becomes too wet, add more browns to balance it out.

Building a basic compost pile is an essential step towards reducing organic waste and recycling valuable nutrients back into the soil. By composting at home, you are not only diverting waste from landfills but also creating a sustainable and eco-friendly solution for your gardening needs. Start your compost pile today and join the organic waste revolution!

Utilizing Compost Bins and Tumblers

Composting has gained significant attention in recent years due to its numerous environmental benefits. By converting organic waste into nutrient-rich compost, we can reduce the amount of waste that ends up in landfills while simultaneously creating a valuable resource for our gardens and plants. One of the most effective ways to embrace the organic waste revolution is by utilizing compost bins and tumblers in every household.

Compost bins and tumblers provide convenient and efficient solutions for managing organic waste. They allow us to compost a wide range of materials, including kitchen scraps, garden waste, and yard trimmings. With these simple tools, anyone can contribute to the reduction of waste and the creation of a healthier environment.

One of the primary advantages of using compost bins and tumblers is their ability to accelerate the decomposition process. These containers create an ideal environment for the organic materials to break down quickly. The controlled aeration and moisture levels inside the bins and tumblers promote the growth of beneficial microorganisms, which speed up the decomposition process. As a result, you can obtain nutrient-rich compost in a matter of weeks or months, depending on the type of bin or tumbler you choose.

Compost bins are perfect for individuals with limited space or those living in urban areas. They are typically made of durable materials such as plastic or wood and come in various sizes to accommodate different needs. These bins are designed to be easy to use, with features like access doors for adding and removing compost, as well as ventilation systems for proper airflow.

On the other hand, tumblers offer the advantage of easy turning and mixing of compost. These cylindrical containers are mounted on a frame or stand, allowing them to be rotated easily. Turning the tumbler a few times a week helps mix the compost materials, ensuring faster decomposition and better aeration. Tumblers are often preferred by individuals who want a more hands-on approach to composting.

No matter which option you choose, compost bins and tumblers are essential tools for every household looking to make a positive impact on the environment. By utilizing these containers, you can divert organic waste from landfills, reduce greenhouse gas emissions, and produce nutrient-rich compost for your plants. Join the organic waste revolution today and embrace the power of composting in your own home.

Indoor Composting

In today's world, where environmental concerns are at the forefront, recycling has become increasingly important. One of the most effective ways to reduce waste and contribute to a sustainable future is through composting. Composting allows us to transform our organic waste into nutrient-rich soil, reducing the need for chemical fertilizers and landfill space. While many people are familiar with traditional outdoor composting methods, indoor composting offers a convenient and practical solution for those living in urban areas or with limited outdoor space.

Indoor composting is a simple and efficient way to recycle your organic waste right in the comfort of your own home. By using a small bin or container, you can collect kitchen scraps such as fruit and vegetable peels, coffee grounds, eggshells, and even paper towels. These everyday materials, which would typically end up in the trash, can be transformed into valuable compost that can nourish your indoor plants or be used in outdoor gardens.

There are several benefits to indoor composting. Firstly, it significantly reduces the amount of waste that goes to landfills, thereby reducing greenhouse gas emissions and environmental pollution. Secondly, it allows you to create a sustainable and natural source of fertilizer for your plants, promoting healthier growth and reducing the need for chemical-based alternatives. Moreover, indoor composting is a year-round activity that can be done regardless of weather conditions, making it a reliable and consistent method of waste management.

Getting started with indoor composting is easy. Begin by selecting a suitable container with a lid to prevent any odors or pests. Make sure it has drainage holes at the bottom to avoid waterlogging. Then, place a layer of shredded newspaper or dried leaves at the bottom to facilitate airflow. As you accumulate your kitchen scraps, remember to add a layer of brown materials, such as shredded paper or dry leaves, to balance the carbon-to-nitrogen ratio. Regularly mix the contents of your compost bin to ensure proper decomposition and prevent any foul odors.

Indoor composting is an excellent way for everyone to contribute to the importance of recycling and environmental sustainability. It allows us to take control of our organic waste, reducing our carbon footprint and creating a valuable resource in the process. By embracing indoor composting, you can make a positive impact on our planet, one kitchen scrap at a time.

Vermicomposting with Worms

One of the most effective ways to recycle organic waste and contribute to a sustainable future is through vermicomposting with worms. This natural process harnesses the power of earthworms to break down organic waste material, turning it into nutrient-rich compost that can be used to nourish plants and gardens. In this subchapter, we will explore the benefits and techniques of vermicomposting, highlighting its importance in the larger context of recycling.

Vermicomposting offers numerous advantages over traditional composting methods. Firstly, it can be done indoors or in small spaces, making it accessible to every household, regardless of their living arrangements. All you need is a suitable container, like a plastic bin or wooden box, some bedding materials, and a population of red worms, which can be easily obtained from local suppliers or online. Secondly, vermicomposting is a relatively quick process, with worms speeding up the decomposition of organic matter by consuming and digesting it. This allows for faster compost production compared to traditional composting methods.

By engaging in vermicomposting, individuals can actively participate in reducing the amount of organic waste that would otherwise end up in landfills. Food scraps, coffee grounds, tea bags, and even paper products like cardboard or newspaper can all be transformed into valuable compost through the work of our little earthworm friends. This not only reduces the environmental impact of waste disposal but also helps to mitigate greenhouse gas emissions that contribute to climate change.

Moreover, the resulting vermicompost is a nutrient-rich fertilizer that promotes healthy plant growth. It improves soil structure, enhances water retention capabilities, and increases nutrient availability, thereby reducing the need for chemical fertilizers. By incorporating vermicompost into gardens, flower beds, or potted plants, individuals can cultivate flourishing vegetation while minimizing their reliance on synthetic and potentially harmful substances.

In conclusion, vermicomposting with worms is a simple yet effective solution for recycling organic waste at the household level. Its accessibility, efficiency, and environmental benefits make it an ideal method for every individual interested in contributing to the organic waste revolution. By practicing vermicomposting, we can divert waste from landfills, reduce greenhouse gas emissions, and create nutrient-rich compost to enrich our gardens. So, let us all join hands and embrace the power of vermicomposting to make a positive impact on our environment and create a greener, more sustainable future for all.

Bokashi Composting

Bokashi Composting: Harnessing the Power of Microbes for Effective Recycling

In the quest for sustainable living, composting has emerged as a vital practice that promotes responsible waste management and environmental stewardship. Among the various composting techniques, Bokashi composting stands out for its incredible efficiency and versatility. This subchapter explores the wonders of Bokashi composting and how it can revolutionize your recycling efforts.

At its core, Bokashi composting is a fermentation process that utilizes a mixture of beneficial microorganisms to break down organic waste. Originating from Japan, this technique has gained popularity worldwide due to its ability to compost almost any organic material, including meat, dairy, and cooked food scraps – items that are typically off-limits in traditional composting methods.

One of the key advantages of Bokashi composting lies in its simplicity and ease of implementation. All you need is a Bokashi bin, airtight containers, a Bokashi starter, and a willingness to embrace the power of beneficial microbes. By layering your organic waste with the Bokashi starter and tightly sealing the bin, you create an anaerobic environment that promotes the growth of microorganisms, which begin breaking down the waste.

The beauty of Bokashi composting is that it can be done indoors, making it an ideal solution for urban dwellers or those lacking outdoor space. With its odorless and efficient process, you can transform your kitchen scraps into a nutrient-rich soil amendment without any unpleasant smells.

Moreover, Bokashi composting offers an excellent opportunity to reduce your carbon footprint. By diverting organic waste from landfills, you not only prevent the emission of harmful greenhouse gases but also contribute to the conservation of valuable landfill space. Additionally, the resulting Bokashi compost can be used to enrich your garden soil, thus closing the loop and completing the recycling cycle.

In summary, Bokashi composting is a game-changer in the world of recycling and organic waste management. Its ability to process a wide range of organic materials, its indoor compatibility, and its positive environmental impact make it an appealing option for every household. Whether you are an urban dweller or an avid gardener, Bokashi composting empowers you to take control of your waste and contribute to the organic waste revolution. By harnessing the power of beneficial microbes, you can turn your kitchen scraps into a valuable resource, reducing your impact on the planet and nourishing your soil in the process. Join the Bokashi revolution and be part of the solution to a greener future!

Chapter 4: Community Composting

In recent years, the importance of recycling has become an increasingly prominent topic of discussion. With growing concerns about climate change, pollution, and the depletion of natural resources, individuals from all walks of life are realizing the need to take action. One area where every household can make a significant impact is through community composting.

Community composting is a simple yet powerful solution that allows individuals to recycle organic waste and transform it into nutrient-rich compost. It involves setting up composting stations in neighborhoods, schools, or even workplaces, where residents can deposit their food scraps, yard waste, and other organic materials. These materials are then mixed and carefully managed to create compost, which can be used to enrich soil, grow healthy plants, and reduce the need for chemical fertilizers.

The benefits of community composting are manifold. Firstly, it reduces the amount of organic waste that ends up in landfills. When organic waste decomposes in landfills, it produces methane, a potent greenhouse gas that contributes to global warming. By diverting this waste to community composting stations, we can significantly reduce greenhouse gas emissions and mitigate the effects of climate change.

Secondly, community composting fosters a sense of community and collaboration. It brings people together to work towards a common goal, creating opportunities for neighbors to connect, share knowledge, and build lasting relationships. Whether you're a seasoned gardener or a complete novice, community composting provides a platform for people of all backgrounds to learn from one another and contribute to the well-being of their environment.

Furthermore, community composting promotes sustainable gardening practices. By using compost instead of synthetic fertilizers, gardeners can improve soil health and fertility, enhance plant growth, and reduce the need for chemical inputs. This not only benefits individual gardeners but also contributes to the overall health of our ecosystems, as it minimizes soil erosion, water pollution, and the depletion of natural resources.

In conclusion, community composting plays a vital role in the larger context of recycling and sustainable living. It empowers individuals to take responsibility for their organic waste and actively participate in the creation of a more environmentally friendly future. By embracing community composting, we can reduce landfill waste, curb greenhouse gas emissions, build stronger communities, and promote sustainable gardening practices. Let us all join hands and make community composting a part of our everyday lives for a greener and healthier planet.

Participating in Local Composting Programs

In today's world, where environmental concerns are at the forefront, it is crucial for every individual to take responsibility for their waste management. One of the most effective ways to address this issue is by participating in local composting programs. These programs play a vital role in the organic waste revolution, offering sustainable and eco-friendly solutions for every household.

Composting is a natural process that converts organic waste into nutrient-rich soil known as compost. By participating in local composting programs, individuals can divert a significant amount of their waste from landfills, where it would otherwise contribute to harmful greenhouse gas emissions. Instead, this waste is transformed into a valuable resource that can be used to enrich soil, nurture plants, and promote a healthy ecosystem.

The importance of recycling cannot be overstated. Through composting, we not only reduce the amount of waste that ends up in landfills but also minimize the need for chemical fertilizers and pesticides. Compost acts as a natural fertilizer, providing essential nutrients to plants, improving soil structure, and enhancing water retention. By participating in local composting programs, individuals can actively contribute to the health of their gardens, local parks, and community green spaces.

Moreover, composting is a simple process that can be easily integrated into every household routine. By separating organic waste such as food scraps, yard trimmings, and leaves from other trash, individuals can create a dedicated composting bin or pile. Local composting programs often provide guidance on best practices, including the correct ratio of greens (nitrogen-rich materials) to browns (carbon-rich materials) for optimal decomposition.

Participating in local composting programs not only benefits the environment but also offers a sense of community engagement. These programs often organize workshops, gatherings, and educational events that bring like-minded individuals together, fostering a spirit of collaboration and shared commitment towards sustainable living. By joining these initiatives, individuals can learn from experts, exchange ideas, and inspire others to adopt composting practices.

In conclusion, participating in local composting programs is a powerful way for EVERY ONE to contribute to the organic waste revolution and the importance of recycling. By diverting waste from landfills and creating nutrient-rich compost, individuals can promote a healthier environment, enhance soil fertility, and foster community engagement. Together, let us embrace the organic waste revolution and make a positive impact on our planet, one compost pile at a time.

Establishing Community Composting Initiatives

In recent years, the importance of recycling and finding sustainable solutions for our waste management has become increasingly evident. As individuals, we have a responsibility to reduce our impact on the environment and contribute to the well-being of our planet. One effective way to achieve this is by establishing community composting initiatives.

Community composting is a grassroots movement that aims to divert organic waste from landfills and transform it into nutrient-rich compost. It involves bringing together individuals, neighborhoods, and even entire cities to collectively compost their organic waste. By doing so, we not only reduce the amount of waste going to landfills but also create a valuable resource that can be used to enrich our soils, promote sustainable agriculture, and reduce the need for chemical fertilizers.

The benefits of community composting initiatives are numerous. Firstly, it helps to reduce greenhouse gas emissions. When organic waste decomposes in landfills, it generates methane, a potent greenhouse gas that contributes to climate change. By composting our organic waste instead, we can significantly reduce these emissions and combat global warming.

Secondly, community composting fosters a sense of community and collaboration. It brings people together, encouraging them to work towards a common goal of sustainability. By engaging in composting initiatives, individuals can connect with their neighbors, share knowledge and resources, and build a stronger, more environmentally conscious community.

Furthermore, community composting initiatives contribute to local food production. The compost produced can be used to nourish community gardens, urban farms, and public green spaces. By using this nutrient-rich compost, we can improve soil health, increase crop yields, and promote sustainable, organic farming practices.

To establish successful community composting initiatives, it is crucial to educate and involve every individual. The book "The Organic Waste Revolution: Composting Solutions for Every Household" is a valuable resource that offers practical guidance on how to initiate and maintain community composting projects. It provides step-by-step instructions, tips on managing composting systems, troubleshooting common issues, and inspiring success stories from existing initiatives.

Ultimately, establishing community composting initiatives is a powerful way for every individual to contribute to the importance of recycling. By working together, we can create a sustainable future and make a positive impact on our environment. So, let us join hands, start composting, and be a part of the organic waste revolution!

Chapter 5: Troubleshooting Common Composting Issues

Dealing with Odor Problems

In our quest for a greener and more sustainable future, recycling and composting play a vital role. As we become more conscious of our waste and its impact on the environment, it is crucial to address the odor problems that can arise from organic waste. In this subchapter, we will explore effective ways to deal with these odor issues and maintain a pleasant living environment while actively participating in the organic waste revolution.

One of the primary reasons for odors in organic waste is improper decomposition. When organic materials such as food scraps and yard waste are not composted correctly, they can produce foul smells. To combat this issue, it is crucial to ensure proper aeration and moisture levels in your compost pile or bin. Turning the compost regularly allows oxygen to reach the decomposing materials, preventing anaerobic decomposition and unpleasant odors. Additionally, maintaining the right moisture balance by periodically adding water or dry materials can help regulate the decomposition process and reduce odors.

Another effective method to tackle odor problems is through the use of odor-absorbing materials. Adding materials such as dry leaves, straw, or sawdust to your compost pile not only provides carbon-rich content but also helps absorb and neutralize odors. These materials act as a natural filter, trapping odorous gases and preventing them from escaping into the air. By incorporating odor-absorbing materials into your composting routine, you can significantly reduce the smell associated with organic waste.

Furthermore, proper waste segregation is crucial in minimizing odor issues. Separating food waste from other types of waste, such as plastic or paper, is essential. By using designated compost bins for organic waste and ensuring proper sealing, you can prevent the smell from spreading throughout your home or living area. Additionally, regularly emptying and cleaning your compost bin can help maintain a fresh and odor-free environment.

Lastly, the importance of recycling cannot be overstated. By actively participating in composting and recycling programs, we contribute to reducing the overall waste that ends up in landfills. Recycling not only helps preserve the environment but also minimizes the potential for odor problems associated with organic waste. By understanding the importance of recycling and actively practicing composting, we can pave the way for a greener future while keeping our living spaces fresh and odor-free.

In conclusion, dealing with odor problems when composting organic waste is essential for creating a pleasant living environment. By ensuring proper decomposition, using odor-absorbing materials, practicing waste segregation, and emphasizing the importance of recycling, we can actively participate in the organic waste revolution while maintaining a fresh and odor-free living space for everyone. Together, let us embrace the power of composting and recycling to create a greener and more sustainable future.

Managing Pests and Rodents

In our journey towards embracing the organic waste revolution and adopting composting solutions for every household, it is crucial to address the issue of managing pests and rodents. As we strive to reduce our environmental impact and promote sustainable practices, it is essential to find effective, eco-friendly ways to handle these challenges.

Pests and rodents can be a nuisance, causing damage to our gardens, homes, and composting systems. However, it is important to approach pest management with a focus on sustainability and the well-being of our environment. Rather than resorting to harmful chemicals and pesticides, let's explore some organic methods to keep pests and rodents at bay.

1. Prevention is Key: The first step in managing pests and rodents is to prevent their entry into our homes and composting areas. Seal any cracks or openings in doors, windows, and walls. Keep compost bins securely closed and use tight-fitting lids. Regularly clean and maintain your composting system to discourage pests from making it their home.

2. Natural Pest Deterrents: Many natural substances act as effective deterrents for pests and rodents. Place mint leaves, garlic, or chili peppers around your composting area to repel pests. Planting marigolds, lavender, or basil around your garden can also help keep pests away. Additionally, consider using essential oils such as neem oil or peppermint oil to repel rodents.

3. Companion Planting: Explore the concept of companion planting, where certain plants are grown together to deter pests. For example, planting marigolds near tomatoes can help repel aphids and other harmful insects. Similarly, growing mint near cabbage plants can deter cabbage worms. By strategically planning your garden and composting area, you can naturally discourage pests and rodents.

4. Biological Pest Control: Another eco-friendly approach is to introduce natural predators of pests into your garden. Encourage beneficial insects like ladybugs, lacewings, and praying mantises by planting flowers they are attracted to. These insects can help control pests such as aphids, caterpillars, and mites.

Remember, managing pests and rodents organically requires patience and persistence. It may take some time to find the right balance and combination of methods that work for your specific situation. By adopting these eco-friendly practices, we can protect our environment, promote sustainable living, and contribute to the organic waste revolution.

Managing pests and rodents is just one aspect of our journey towards composting solutions for every household. By incorporating these practices into our daily lives, we can reap the benefits of a healthier environment, thriving gardens, and sustainable waste management. Let's continue to prioritize the importance of recycling and make a positive impact on our planet, one compost bin at a time!

Addressing Excessive Moisture or Dryness

Moisture plays a vital role in the composting process. It helps break down organic waste and creates an ideal environment for beneficial microorganisms to thrive. However, excessive moisture or dryness can hinder the composting process and lead to unpleasant odors or a lack of decomposition. In this subchapter, we will explore how to address excessive moisture or dryness in your compost pile, ensuring a successful composting journey and contributing to the organic waste revolution.

Excessive moisture is often caused by adding too many wet or high-moisture materials, such as food scraps or fresh grass clippings, without balancing them with dry or carbon-rich materials like leaves or shredded paper. To combat this issue, it is essential to maintain the right moisture balance. A simple way to achieve this is by layering your compost pile. Alternate layers of wet and dry materials, ensuring each layer is no more than 2-3 inches thick. This will allow for proper aeration and prevent excess moisture buildup.

In cases of excessive moisture, turning the compost pile regularly can help promote airflow and reduce moisture levels. Additionally, adding dry materials like straw or sawdust can help absorb excess moisture. If the compost pile remains too wet, consider covering it with a tarp or moving it to a more sheltered area to protect it from rain or excessive watering.

On the other hand, dryness can slow down the composting process as well. If your compost pile is too dry, it may take longer for organic materials to decompose and break down. To address this, ensure that you regularly water your compost pile, especially during dry periods. Use a garden hose or a watering can to moisten the materials evenly. However, be cautious not to overwater, as this can lead to excessive moisture and anaerobic conditions.

In conclusion, addressing excessive moisture or dryness in your compost pile is crucial for successful composting. By maintaining the right moisture balance through layering, turning, and watering, you can create an ideal environment for organic waste decomposition. Remember, composting is a key component of the organic waste revolution and contributes to the importance of recycling. Together, we can make a positive impact on the environment by reducing waste and creating nutrient-rich compost for our gardens.

Moisture control is a crucial aspect of successful composting. Whether you live in a humid climate or a dry region, finding the right balance of moisture in your compost pile is essential for the breakdown of organic waste. In this subchapter, we will explore the different ways to address excessive moisture or dryness in your composting process.

Excessive moisture is a common issue that can lead to a smelly, anaerobic compost pile. If your compost is too wet, it can become compacted and lack the necessary airflow for decomposition. To address this problem, there are several steps you can take. First, make sure your compost bin or pile is in a well-drained area. If it is sitting in a low-lying spot that collects water, consider moving it to a higher location. Additionally, you can add dry, carbon-rich materials such as straw, dried leaves, or shredded newspaper to absorb excess moisture. Turning the pile regularly will also help to increase airflow and reduce moisture levels.

On the other hand, if your compost pile is too dry, it may take longer for the organic waste to decompose. In this case, you can add moisture by watering the pile. Use a hose or watering can to evenly distribute water throughout the pile. Be careful not to overwater, as this can lead to anaerobic conditions. It is recommended to add water gradually and monitor the moisture levels to achieve the ideal balance. Another option is to cover the pile with a tarp or plastic sheet to help retain moisture and create a more humid environment.

Regardless of whether you are dealing with excessive moisture or dryness, monitoring the moisture levels of your compost pile is crucial. A simple way to check the moisture content is to squeeze a handful of compost. If water drips out, it is too wet, and if it crumbles, it is too dry. A well-moistened pile should feel like a damp sponge.

By addressing excessive moisture or dryness in your composting process, you are ensuring optimal conditions for the breakdown of organic waste. This not only speeds up the composting process but also helps to prevent odors and create nutrient-rich compost for your garden. Remember, maintaining the right moisture balance is just one of the many steps towards becoming a more sustainable and environmentally conscious individual.

In conclusion, understanding how to address excessive moisture or dryness in your composting process is essential for successful recycling of organic waste. By implementing the tips and techniques mentioned in this subchapter, you can create a compost pile that thrives and contributes to the organic waste revolution.

Troubleshooting Slow or Incomplete Composting

Composting is an excellent way to reduce organic waste and create nutrient-rich soil for your garden. However, sometimes the composting process may become slow or incomplete, leaving you with a pile of rotting waste instead of rich, dark compost. Don't worry, though - with a little troubleshooting, you can get your compost back on track and continue contributing to the organic waste revolution.

One common reason for slow or incomplete composting is a lack of proper balance in your compost pile. Composting requires a mix of green and brown materials. Greens, such as grass clippings and kitchen scraps, provide nitrogen, while browns, like dry leaves and twigs, provide carbon. If your compost seems slow, it may be because you have too much of one type of material. Adjust the balance by adding more greens or browns accordingly. Remember to mix the materials well to ensure proper decomposition.

Another reason for slow composting is inadequate moisture. Compost needs to be moist, but not overly wet, for the bacteria and organisms to break down the organic matter effectively. If your compost is dry, try adding water to the pile and mixing it thoroughly. On the other hand, if it's too wet, add some dry browns to absorb the excess moisture and improve aeration.

Insufficient aeration can also hinder the composting process. Oxygen is vital for the decomposition process, and without it, your compost may become anaerobic, resulting in a foul smell. To improve aeration, turn the pile regularly using a pitchfork or compost turner. This will introduce fresh air and help the microorganisms thrive, thus speeding up decomposition.

Additionally, the size of your compost pile can impact its speed and completeness. A smaller pile may not generate enough heat to break down the organic matter efficiently. Consider increasing the size of your compost pile to enhance the heat-generating process and promote faster decomposition.

Remember that patience is key when troubleshooting slow or incomplete composting. Composting is a natural process that takes time, and occasionally, adjustments are needed to achieve the desired results. By maintaining a proper balance of materials, providing adequate moisture and aeration, and being mindful of pile size, you can troubleshoot any issues and continue contributing to the organic waste revolution.

In conclusion, troubleshooting slow or incomplete composting may require adjustments to the balance, moisture, aeration, and pile size. By following these guidelines, you can overcome common obstacles and ensure a successful composting experience. Join the organic waste revolution and make a positive impact on the environment by recycling your organic waste through composting.

Composting is a great way to reduce waste and contribute to a more sustainable future. However, sometimes the process may not go as smoothly as we hope. If you find that your compost is slow to decompose or is not breaking down completely, don't worry! There are several common issues that can be easily addressed to ensure successful composting.

One of the main reasons for slow or incomplete composting is the lack of proper balance in the compost pile. Composting requires a good mix of green and brown materials. Green materials, such as kitchen scraps and fresh grass clippings, provide nitrogen, while brown materials, such as dry leaves and woody trimmings, provide carbon. If your compost seems to be taking longer than usual, it may be because there is an imbalance in the ratio of these materials. Try adding more brown materials to the pile to provide the necessary carbon, and make sure to turn the pile regularly to mix everything together.

Another common issue is moisture content. Composting organisms need moisture to thrive, but excessive moisture can lead to a slimy, anaerobic environment that slows down decomposition. On the other hand, if your compost is too dry, it will take much longer to break down. To maintain the ideal moisture level, aim for a damp sponge consistency. If your compost is too wet, add dry materials like shredded newspaper or straw. If it's too dry, sprinkle some water while turning the pile.

Temperature is also a crucial factor in composting. The microorganisms responsible for breaking down organic matter work best in warm conditions. If your compost is not heating up, it may be due to insufficient nitrogen or lack of aeration. Ensure that you have enough green materials in your pile and turn it regularly to introduce oxygen. If the compost still doesn't heat up, consider adding a compost activator or inoculant to introduce more beneficial microorganisms.

Lastly, keep in mind that some materials are simply not suitable for composting. Avoid adding meat, dairy, or oily foods, as they can attract pests and slow down the decomposition process. Similarly, large branches or woody materials will take much longer to break down and may need to be chipped or shredded before adding them to the pile.

By troubleshooting these common issues, you can ensure that your composting process is both efficient and effective. Remember, composting is an essential part of the recycling process, turning organic waste into nutrient-rich soil amendment for your garden. So don't get discouraged if you encounter some challenges along the way – with a little bit of knowledge and effort, you can become a composting pro and contribute to the organic waste revolution!

Chapter 6: Utilizing Compost in Your Household

Understanding the Benefits of Compost

Composting is a simple yet effective way to recycle organic waste and play a vital role in the larger picture of environmental sustainability. In this subchapter, titled "Understanding the Benefits of Compost," we will delve into the various advantages of composting, addressing everyone from seasoned environmentalists to those just starting their journey towards a greener lifestyle.

1. Environmental Impact: Composting significantly reduces the amount of waste that ends up in landfills, where it decomposes and releases harmful gases such as methane. By diverting this waste to compost piles, we can minimize greenhouse gas emissions and alleviate the strain on landfill capacity, contributing to a healthier planet.

2. Soil Enrichment: Compost is often referred to as "black gold" due to its incredible ability to enhance soil fertility. When added to gardens, yards, and farms, compost enriches the soil with essential nutrients, improves its structure, and enhances moisture retention. This leads to healthier plants, increased crop yields, and a more sustainable and resilient food system.

3. Waste Reduction: Composting allows us to reduce the amount of waste we produce at home. By diverting food scraps, yard trimmings, and other organic materials from the trash, we can significantly decrease the volume of waste that goes to landfills. This not only

reduces the need for waste management infrastructure but also saves valuable resources.

4. Cost Savings: Composting can lead to substantial cost savings for households and communities. By composting at home, individuals can reduce the amount of garbage they produce, potentially lowering their waste management fees. Moreover, municipalities can save on waste collection and disposal expenses, which can be redirected towards other important community initiatives.

5. Water Conservation: Compost helps retain moisture in the soil, reducing the need for excessive watering. By improving soil structure and water-holding capacity, compost helps plants thrive even during dry spells. This contributes to water conservation efforts and helps mitigate the impact of droughts on local ecosystems.

6. Closing the Loop: Composting completes the cycle of organic matter. By returning nutrients from food scraps and other organic waste back to the soil, we close the loop in the natural cycle of life, fostering a more sustainable and self-reliant ecosystem.

In conclusion, understanding the benefits of composting is crucial for everyone, regardless of their level of knowledge or interest in environmental sustainability. From reducing landfill waste and improving soil fertility to conserving water and saving money, composting offers a multitude of advantages for individuals, communities, and the planet as a whole. By embracing composting as a simple yet powerful solution, we can all contribute to the organic waste revolution and pave the way towards a greener and more sustainable future.

Composting is a powerful tool that allows us to harness the natural process of decomposition to create nutrient-rich soil. In this subchapter, we will explore the various benefits of composting and why it is an essential practice for every household. Whether you are a gardening enthusiast, an eco-conscious individual, or simply someone interested in reducing waste, understanding the benefits of composting will inspire you to take action.

First and foremost, composting is a sustainable way to manage organic waste. By diverting food scraps and yard waste from landfills, we can significantly reduce methane gas emissions, a potent greenhouse gas that contributes to climate change. Additionally, composting reduces the need for chemical fertilizers, pesticides, and herbicides, which can harm the environment and our health. By recycling organic waste into nutrient-rich compost, we create a closed-loop system that supports healthy soil and plant growth.

Compost is often referred to as "black gold" due to its numerous benefits for gardening and agriculture. When added to soil, compost improves its structure, allowing for better water retention and drainage. It also enhances soil fertility by providing essential nutrients and beneficial microorganisms that promote healthy plant growth. Compost acts as a natural fertilizer, reducing the need for synthetic alternatives that can contaminate groundwater and harm beneficial insects.

Another advantage of composting is its ability to remediate contaminated soil. By adding compost to polluted areas, we can help break down and neutralize harmful chemicals, heavy metals, and pesticides, restoring the soil's health and creating a safer environment for plants and animals.

Beyond its environmental benefits, composting also saves money. By reducing the amount of waste sent to landfills, municipalities can save on disposal costs, ultimately leading to potential savings for taxpayers. Moreover, by producing your own compost, you eliminate the need to purchase expensive fertilizers, contributing to significant long-term savings for avid gardeners and farmers.

In conclusion, composting is a simple and effective way to address the growing problem of organic waste while providing numerous benefits for both the environment and our wallets. From improving soil health and fertility to reducing greenhouse gas emissions and saving money, composting is a practice that should be embraced by everyone. By understanding the benefits of composting, we can take small steps towards a more sustainable future and contribute to the organic waste revolution.

Using Compost in the Garden

Composting is more than just a way to reduce waste; it is a powerful tool that can transform your garden into a vibrant and thriving ecosystem. In this subchapter, we will explore the various benefits of using compost in the garden and how it contributes to the importance of recycling.

Compost is a nutrient-rich organic material that is created through the natural decomposition of organic waste such as food scraps, yard trimmings, and leaves. By diverting these materials from the landfill and turning them into compost, we can reduce greenhouse gas emissions and conserve valuable landfill space. But the benefits of composting extend far beyond waste management.

One of the key advantages of using compost in the garden is its ability to improve soil health. Compost is packed with essential nutrients that plants need to grow and thrive. When incorporated into the soil, it enriches it with organic matter, enhances its structure, and increases its water-holding capacity. This means healthier plants that are more resistant to diseases and pests, and better yields for your vegetable garden.

Compost also acts as a natural fertilizer, slowly releasing nutrients over time. Unlike synthetic fertilizers, which can harm beneficial soil organisms and contaminate water sources, compost feeds the soil in a sustainable and environmentally friendly way. It helps maintain a balanced pH level, promotes beneficial microbial activity, and improves soil structure, which is particularly important in areas with heavy clay or sandy soils.

Furthermore, using compost in the garden promotes biodiversity and supports a wide range of beneficial insects, earthworms, and microorganisms. These organisms play crucial roles in breaking down organic matter, improving soil aeration, and suppressing harmful pests. By creating a healthy and diverse soil ecosystem, composting contributes to the overall health of your garden and reduces the need for harmful pesticides.

In conclusion, using compost in the garden not only helps in the recycling process but also provides numerous benefits for your plants and the environment. It improves soil health, increases plant resilience, and supports biodiversity. By harnessing the power of composting, we can turn our organic waste into a valuable resource and contribute to a more sustainable and eco-friendly world. So, why not start composting today and unlock the potential of your garden?

Composting is a natural process that converts organic waste into nutrient-rich compost, which can be used to improve the health and vitality of gardens. In this subchapter, we will explore the many benefits of using compost in the garden and how it contributes to the importance of recycling.

Compost is often referred to as "black gold" for gardeners, and for good reason. It is a valuable soil amendment that helps retain moisture, improves soil structure, and enhances nutrient availability. Whether you have a small urban garden or a sprawling backyard, incorporating compost into your gardening routine can have a significant positive impact on the overall health of your plants.

One of the key benefits of using compost is its ability to retain moisture in the soil. The organic matter in compost acts like a sponge, absorbing water and releasing it slowly over time. This helps to prevent water runoff and reduces the need for frequent watering, especially during dry periods. Additionally, compost helps to improve soil structure, making it easier for plant roots to penetrate and access water and nutrients.

Another advantage of using compost is its ability to enrich the soil with essential nutrients. Compost is rich in organic matter, which serves as a food source for beneficial soil organisms such as earthworms and bacteria. These organisms break down the organic matter, releasing nutrients in a form that plants can readily absorb. By replenishing the soil with compost, you are providing a continuous supply of nutrients to your plants, reducing the need for synthetic fertilizers.

In addition to its benefits for plant growth, using compost in the garden also contributes to the importance of recycling. Organic waste, such as food scraps and yard trimmings, make up a significant portion of household waste that would otherwise end up in landfills. By composting these materials, you are diverting them from the waste stream and turning them into a valuable resource. In doing so, you are not only reducing the amount of waste that goes to landfills but also minimizing greenhouse gas emissions associated with waste decomposition.

In conclusion, using compost in the garden is a win-win situation. It improves soil health, conserves water, reduces the need for synthetic fertilizers, and contributes to the importance of recycling. Whether you are a seasoned gardener or just starting out, incorporating compost into your gardening routine is a simple and effective way to promote sustainable practices while reaping the rewards of a thriving garden.

Improving Soil Quality with Compost

Composting is a powerful tool that can revolutionize the way we view organic waste. In this subchapter, we will explore how composting can enhance soil quality and its significance in the broader context of recycling. Whether you are a gardening enthusiast, an environmentally conscious individual, or simply interested in sustainable practices, this information is relevant to everyone.

Compost, often referred to as "black gold," is the result of the decomposition of organic materials such as kitchen scraps, yard waste, and plant residues. This natural process breaks down these materials into a nutrient-rich soil amendment that can greatly benefit our gardens, lawns, and agricultural fields.

One of the primary advantages of using compost is its ability to enhance soil quality. Compost improves soil structure, making it more resistant to erosion and compaction. It also increases the soil's water-holding capacity, reducing the need for excessive irrigation. The organic matter in compost promotes beneficial microbial activity, creating a healthier soil ecosystem that supports plant growth and suppresses diseases.

Furthermore, compost acts as a slow-release fertilizer, providing essential nutrients to plants over an extended period. Unlike synthetic fertilizers, compost nourishes plants without the risk of nutrient runoff, which can contaminate water sources and harm aquatic life. By utilizing compost, we can reduce our reliance on chemical fertilizers and minimize the environmental impact associated with their production and use.

In the broader context of recycling, composting is a crucial component of the circular economy. By diverting organic waste from landfills and transforming it into valuable compost, we close the loop on the waste stream. This reduces greenhouse gas emissions, as organic materials in landfills release methane, a potent greenhouse gas. Composting also helps to conserve landfill space and reduces the need for costly waste management practices.

Regardless of your background or interests, understanding the importance of recycling is key. Recycling is about more than just separating plastics and paper; it encompasses the entire waste management hierarchy, including composting. By embracing composting as a solution for organic waste, we contribute to the larger goal of creating a sustainable future.

In conclusion, composting is a powerful tool that improves soil quality and plays a vital role in the broader context of recycling. By utilizing compost, we can enhance our gardens, reduce water usage, and minimize reliance on synthetic fertilizers. Additionally, composting diverts organic waste from landfills, reducing greenhouse gas emissions and conserving valuable resources. Embracing composting as a solution for organic waste is a simple yet effective way for everyone to contribute to a more sustainable world.

Composting is a simple and effective way to recycle organic waste and improve soil quality. In this subchapter, we will delve into the importance of recycling and how composting can be a solution for every household.

Recycling is crucial in today's world as it helps reduce the amount of waste that ends up in landfills, conserves resources, and minimizes pollution. However, recycling is not limited to just plastic, paper, and metal. Organic waste, such as kitchen scraps and yard trimmings, can also be recycled through composting.

Composting is the process of breaking down organic waste into nutrient-rich soil, known as compost. This natural fertilizer is teeming with beneficial microorganisms, essential nutrients, and organic matter. By adding compost to soil, you can enhance its fertility, structure, and moisture-holding capacity.

One of the key benefits of composting is its ability to improve soil quality. When compost is added to sandy soil, it helps bind particles together and improve water retention. In contrast, when added to clay soil, compost improves drainage and aeration, preventing it from becoming compacted. Compost also supplies essential nutrients to plants, promoting healthy growth and increasing their resistance to pests and diseases.

Composting is not only beneficial for large-scale agricultural practices but also for every household. Whether you have a spacious backyard or a small balcony, there are composting solutions available for everyone. You can choose from traditional compost bins, worm composting systems, or even indoor composters that fit neatly under your sink.

By composting at home, you are not only reducing your carbon footprint but also saving money. Instead of purchasing chemical fertilizers and soil amendments, you can create your own nutrient-rich compost using kitchen scraps, yard waste, and other organic materials. This not only saves you money but also eliminates the need for harmful synthetic chemicals in your garden, promoting a healthier environment for you and your family.

In conclusion, recycling is of utmost importance for the well-being of our planet. Composting is a powerful tool that allows us to recycle organic waste and improve soil quality. Whether you are an avid gardener or simply want to make a positive impact, composting is a solution that is accessible to everyone. Start composting today and join the organic waste revolution!

Compost Tea and Other Liquid Compost Applications

In this subchapter, we will explore the fascinating world of compost tea and other liquid compost applications. Composting has long been recognized as an effective way to reduce organic waste and create nutrient-rich soil amendments. However, traditional composting methods can sometimes be time-consuming and require significant space. That's where liquid compost applications come in, offering a convenient and efficient alternative for recycling organic waste.

Compost tea is a liquid extract derived from compost that contains a multitude of beneficial microorganisms. These microorganisms help break down organic matter, release nutrients, and suppress harmful pathogens in the soil. When applied to plants, compost tea acts as a natural fertilizer, enhancing plant growth and overall health.

One of the key advantages of compost tea is its versatility. It can be used in various ways, such as foliar sprays, soil drenches, or even as a seed soak. Foliar sprays involve spraying the liquid compost directly onto plant leaves, allowing the nutrients to be absorbed through the leaves' stomata. Soil drenches, on the other hand, involve applying the compost tea directly to the soil, where it can penetrate the root zone and nourish plants from below. Both methods provide plants with the essential nutrients they need to thrive.

Liquid compost applications extend beyond compost tea, with other innovative options available. For example, compost leachate is the liquid that drains from a compost pile. Although it may initially seem like a waste product, compost leachate can be collected and utilized as a liquid fertilizer. It contains many of the same beneficial microorganisms and nutrients found in compost tea, making it an excellent option for nourishing plants.

The importance of recycling organic waste cannot be overstated. By diverting food scraps, yard trimmings, and other organic materials from landfills, we can significantly reduce greenhouse gas emissions and create a more sustainable future. Compost tea and other liquid compost applications provide a practical solution for every household to participate in this organic waste revolution.

Whether you are an avid gardener, a horticulture enthusiast, or simply someone interested in sustainable practices, liquid compost applications offer a world of possibilities. With just a few simple steps, you can transform your organic waste into a nutrient-rich elixir that promotes healthy plant growth and supports the environment.

So, join the movement today and discover the wonders of compost tea and other liquid compost applications. Together, we can make a difference in our communities and contribute to a greener, healthier planet for future generations.

Composting is known to be one of the most effective ways to reduce organic waste and create nutrient-rich soil for gardening and farming. But did you know that there are other innovative ways to utilize compost? In this subchapter, we will explore the benefits and applications of compost tea and other liquid compost solutions.

Compost tea is essentially a liquid extract made from compost that contains a concentrated mix of beneficial microorganisms, nutrients, and organic matter. This nutrient-rich liquid can be used as a foliar spray or soil drench to promote plant health and enhance soil fertility.

The process of making compost tea involves steeping compost in water and aerating it to encourage the growth of beneficial bacteria, fungi, and other microorganisms.

One of the main advantages of compost tea is its ability to boost plant growth and suppress diseases. When sprayed onto plants, it acts as a natural fertilizer, providing essential nutrients and stimulating root development. The beneficial microorganisms present in compost tea can also help protect plants from harmful pathogens by outcompeting them for resources and creating a healthier soil environment.

Not only is compost tea beneficial for plants, but it also helps in the overall recycling process. By using liquid compost solutions, we can make the most out of our organic waste and reduce the amount of waste sent to landfills. This directly contributes to the importance of recycling and highlights the potential of organic waste in creating sustainable solutions for every household.

Liquid compost applications go beyond just compost tea. Other liquid compost solutions include compost extracts, compost leachate, and compost filtrates. Each of these variations offers unique benefits and can be tailored to suit specific gardening or farming needs.

In conclusion, compost tea and other liquid compost applications are valuable tools in the organic waste revolution. By utilizing these liquid solutions, we can harness the nutrient-rich properties of compost to improve plant health, enhance soil fertility, and contribute to the importance of recycling. Whether you are an avid gardener or a commercial farmer, incorporating liquid compost applications into your practices can lead to more sustainable and eco-friendly outcomes. Join the organic waste revolution and explore the vast potential of compost tea and other liquid compost solutions today.

Composting for Indoor Plants

In our modern world, where environmental concerns are becoming increasingly important, recycling and composting have emerged as crucial solutions for sustainable living. Composting, in particular, is a valuable practice that not only reduces waste but also nourishes the soil, making it an ideal solution for indoor plant lovers. In this subchapter, we will explore the benefits of composting for indoor plants and how you can easily implement it in your own home.

Indoor plants are not only aesthetically pleasing but also provide numerous health benefits, such as purifying the air and reducing stress. However, to ensure their optimal growth and longevity, it is essential to provide them with nutrient-rich soil. This is where composting comes into play. By composting organic waste, such as kitchen scraps, coffee grounds, and plant trimmings, you can create a nutrient-dense soil amendment that will boost the health and vitality of your indoor plants.

Composting for indoor plants offers several advantages. Firstly, it helps to reduce the amount of waste that ends up in landfills, thus minimizing the production of harmful greenhouse gases. Secondly, it allows you to take an active role in recycling and reusing organic materials, turning them into a valuable resource for your plants. Additionally, composting can save you money by eliminating the need to purchase expensive soil amendments or fertilizers.

To start composting for your indoor plants, you will need a compost bin or container that fits your space. It can be as simple as a plastic bin with a lid or a specialized composting system. Next, collect organic waste from your kitchen, such as fruit and vegetable scraps, tea bags, and eggshells. Avoid using meat or dairy products, as they can attract pests. Layer the organic waste with dry materials like shredded newspaper, dried leaves, or sawdust to maintain a proper balance of carbon and nitrogen.

Remember to regularly turn the compost pile to ensure proper aeration and decomposition. Over time, the organic waste will break down into a dark, crumbly substance known as "black gold" – your homemade compost. Mix this nutrient-rich compost into the soil of your indoor plants to provide them with essential nutrients and improve moisture retention.

Composting for indoor plants is a simple and sustainable practice that anyone can incorporate into their daily routine. By recycling your organic waste and creating your own compost, you not only contribute to the importance of recycling but also provide your indoor plants with the nourishment they need to thrive. So, let's join the organic waste revolution and start composting for a greener, healthier future!

Indoor plants not only add a touch of beauty and vibrancy to our living spaces but also provide numerous health benefits, such as purifying the air and reducing stress. However, in order to keep these plants thriving and healthy, it is essential to provide them with the right nutrients. This is where composting comes into play. Composting for indoor plants is a sustainable and efficient way to nourish your beloved green companions while also contributing to the importance of recycling.

Composting is the process of breaking down organic waste into nutrient-rich soil, known as compost. By recycling kitchen scraps, such as fruit and vegetable peels, coffee grounds, and eggshells, you can create a valuable resource that will enrich the soil and promote the growth of your indoor plants. The benefits of composting extend far beyond just the environmental aspect; it also helps you reduce waste, save money on fertilizers, and create a more sustainable lifestyle.

To start composting for your indoor plants, you will need a compost bin or container. This can be as simple as a small plastic bin with a lid or a specialized composter designed for indoor use. Next, gather your organic waste materials, ensuring they are free from any chemicals or non-biodegradable substances. Chop or shred the waste into smaller pieces to speed up the decomposition process. Layer the waste in the compost bin, alternating between dry materials like leaves or shredded newspaper and wet materials like food scraps and coffee grounds. You can also add a handful of soil to introduce beneficial microbes that aid in the composting process.

It is important to maintain the right balance of moisture and air in your compost. Keep the compost moist, but not overly wet, by adding water as needed. Additionally, turn the compost regularly to encourage aeration and prevent the formation of odors or mold. Within a few weeks or months, depending on the size of your compost pile and the conditions, you will have a nutrient-rich compost ready to be used in your indoor plants.

By composting for your indoor plants, you are not only reducing waste and promoting sustainability but also providing your plants with a natural and chemical-free source of nutrients. So, let's join the organic waste revolution and start composting for healthier, happier indoor plants!

Creative Uses for Compost in Everyday Life

Composting is a simple yet effective way to recycle organic waste and reduce our carbon footprint. By transforming kitchen scraps, yard trimmings, and other biodegradable materials into nutrient-rich compost, we can create a sustainable cycle that benefits both our environment and our everyday lives. In this subchapter, we will explore some creative uses for compost in our daily routines, highlighting the importance of recycling and its impact on our planet.

1. Nourishing our gardens: Compost is often referred to as "black gold" for its ability to enrich the soil. By adding compost to our gardens, we can improve its structure, enhance water retention, and provide essential nutrients to our plants. Whether it's a small vegetable patch or a flower bed, compost acts as a natural fertilizer, promoting healthy growth and reducing the need for synthetic chemicals.

2. Houseplant revitalizer: Just as compost benefits outdoor plants, it can also work wonders for our indoor greenery. Mixing compost into potting soil or using it as a top dressing can provide a natural boost of nutrients, keeping our houseplants vibrant and thriving. Additionally, compost helps retain moisture, reducing the frequency of watering and ensuring healthier root systems.

3. Natural pest control: Compost can be used as a safe and eco-friendly alternative to chemical pesticides. By creating a compost tea, a liquid fertilizer made by steeping compost in water, we can deter pests and boost our plants' resistance to diseases. This natural pest control method not only protects our plants but also eliminates the harmful effects of chemical sprays on our environment.

4. Compost as mulch: Mulching is a common practice in gardening to suppress weeds, retain moisture, and regulate soil temperature. Compost can be used as an organic mulch, applied around plants or spread over garden beds. Not only does compost provide these benefits, but it also decomposes over time, adding valuable nutrients to the soil.

5. Composting for crafts: Get creative with compost by incorporating it into various arts and crafts projects. Use compost in handmade paper-making, creating unique textures and adding an eco-friendly touch. Compost can also be used as a natural dye, giving fabrics a rustic, earthy appearance.

By exploring these creative uses for compost in our everyday lives, we can highlight the importance of recycling and inspire others to join the organic waste revolution. Recycling our organic waste through composting not only reduces landfill waste but also provides us with a valuable resource that can improve our gardens, nourish our plants, and promote a healthier environment. Let's embrace the power of compost and make a positive impact on our planet, one recycled banana peel at a time.

Composting has become an integral part of the organic waste revolution and is gaining popularity among households worldwide. Not only does it provide an eco-friendly solution for disposing of organic waste, but it also offers numerous benefits for our gardens, plants, and everyday life. In this subchapter, we will explore the creative uses of compost in our daily routines, showcasing its importance in recycling and sustainable living.

1. Nourishing Your Garden: Compost is a natural fertilizer and soil conditioner that enriches the nutrients in the soil. By incorporating compost into your garden beds, you are providing essential organic matter that promotes healthy plant growth, improves water retention, and reduces the need for chemical fertilizers. The result? Vibrant flowers, thriving vegetables, and a lush green landscape.

2. Indoor Plant Care: Compost can also be used for houseplants, creating a nutrient-rich potting mix. Mix compost with regular potting soil to enhance plant growth, improve moisture retention, and increase the overall health of your indoor plants. By using compost, you reduce the reliance on synthetic fertilizers and create a sustainable environment for your houseplants.

3. Composting Tea: Compost tea is an excellent way to utilize the nutrients from your compost. By steeping compost in water, you create a nutrient-dense liquid that can be used as a natural and organic fertilizer for your plants. Simply strain the tea and apply it to your garden or indoor plants. This homemade fertilizer will boost plant growth and improve the overall health of your greenery.

4. Natural Pest Control: Compost can act as a natural deterrent to pests in your garden. By spreading a thin layer of compost around your plants, you create a barrier that repels insects and pests, reducing the need for harmful chemical pesticides. Additionally, the healthy soil ecosystem created by compost attracts beneficial insects that prey on pests, creating a balanced and sustainable garden environment.

5. Compost for Crafts: Compost can be utilized creatively in various arts and crafts projects. From creating handmade paper to making

natural dyes, compost offers a unique and sustainable resource. The organic matter in compost can also be used for creating natural pigments or as a medium for sculpting. By exploring the artistic potential of compost, you contribute to the importance of recycling and showcase the versatility of organic waste.

In conclusion, composting is not only a sustainable and eco-friendly way to dispose of organic waste but also offers a wide range of creative uses in our everyday lives. From nourishing our gardens to providing natural pest control, compost is a valuable resource. By incorporating compost into our routines, we contribute to the importance of recycling, reducing waste, and living a more sustainable lifestyle. So, let's embrace the organic waste revolution and unlock the creative potential of compost for a greener and healthier future.

Chapter 7: Taking Organic Waste Recycling Further

Composting on a Larger Scale

When it comes to recycling, one of the most important aspects that often gets overlooked is organic waste. The staggering amount of food scraps, yard trimmings, and other organic materials that end up in landfills is not only wasteful but also harmful to the environment. That is where composting comes in, and it is not just limited to individual households. Composting on a larger scale is an essential part of the organic waste revolution.

Composting on a larger scale refers to the practice of composting in community settings, institutions, and even industrial facilities. It involves the collection and processing of organic waste from a broader range of sources. By taking composting to a larger scale, we can significantly reduce the amount of organic waste that ends up in landfills and instead turn it into a valuable resource.

One of the primary benefits of composting on a larger scale is the significant reduction of greenhouse gas emissions. When organic waste decomposes in landfills, it releases methane, a potent greenhouse gas that contributes to climate change. By diverting this waste to composting facilities, the organic materials can be properly managed and turned into nutrient-rich soil amendments. This not only reduces methane emissions but also helps to sequester carbon in the soil, mitigating climate change impacts.

Furthermore, composting on a larger scale can create local economic opportunities. Community composting programs can provide employment and training for individuals interested in sustainable waste management. Additionally, the resulting compost can be sold to farmers, gardeners, and landscapers, stimulating the local economy and reducing the need for chemical fertilizers.

To implement composting on a larger scale, collaboration and education are key. Municipalities, businesses, and community organizations must work together to establish composting facilities and develop effective collection systems. Public awareness campaigns and educational programs can help individuals understand the importance of separating organic waste and participating in community composting initiatives.

In conclusion, composting on a larger scale is a crucial component of the organic waste revolution. By diverting organic waste from landfills and turning it into compost, we can reduce greenhouse gas emissions, improve soil health, and stimulate local economies. It is an opportunity for everyone to contribute to the importance of recycling and make a positive impact on the environment.

The importance of recycling is something that resonates with everyone, regardless of their background or lifestyle. As we become increasingly aware of the impact of our actions on the environment, finding sustainable solutions to manage organic waste has become a pressing issue. Composting, in particular, has emerged as a practical and eco-friendly method to tackle this problem. While composting at home is a great starting point, it is essential to explore larger-scale composting solutions to maximize the benefits.

Composting on a larger scale involves managing organic waste from a community or even an entire city. By centralizing the composting process, we can harness the collective power of many households and institutions, significantly increasing the impact of recycling efforts. Large-scale composting facilities provide an efficient way to process a substantial amount of organic waste, diverting it from landfills and transforming it into valuable compost.

One of the key advantages of large-scale composting is its ability to handle a wider range of organic materials. While home composting may be limited to kitchen scraps and yard waste, larger facilities can accept additional inputs such as food waste from restaurants, grocery stores, and even agricultural waste. This comprehensive approach not only reduces the amount of waste sent to landfills but also helps to close the loop in the food production cycle, creating a sustainable and circular system.

In addition to waste diversion, large-scale composting offers numerous environmental benefits. The organic matter, when processed correctly, produces nutrient-rich compost that can be used to enhance soil fertility, reduce the need for chemical fertilizers, and promote healthier plant growth. Moreover, composting on a larger scale reduces greenhouse gas emissions that would have otherwise been generated from decomposing organic waste in landfills.

Implementing large-scale composting solutions requires collaboration between government bodies, waste management agencies, and the community. It involves setting up infrastructure, educating the public about the importance of composting, and establishing efficient collection systems. However, the long-term benefits far outweigh the initial investment and efforts required.

By expanding composting efforts to a larger scale, we can make a significant positive impact on our environment. It is a vital step towards creating a sustainable future, where organic waste is seen as a valuable resource rather than a burden. Together, we can embrace the organic waste revolution and build a greener and healthier planet for generations to come.

Community Initiatives for Organic Waste Recycling

In recent times, there has been a growing awareness about the importance of recycling and its impact on the environment. One crucial aspect of recycling that is gaining traction is organic waste recycling. As individuals and communities become more conscious of their ecological footprint, community initiatives for organic waste recycling have become increasingly popular.

Organic waste refers to any biodegradable waste material that comes from plant or animal sources. This includes food scraps, yard waste, paper, and wood products. Instead of sending these materials to landfills, where they emit harmful greenhouse gases, communities are now taking proactive steps to recycle them through various initiatives.

One of the most prominent community initiatives for organic waste recycling is composting. Composting is a natural process that decomposes organic waste into nutrient-rich soil, which can then be used in gardens, farms, and landscaping. By composting their organic waste, communities not only reduce the amount of waste sent to landfills but also create a valuable resource that enriches the soil and promotes sustainable agriculture.

Many communities have established composting programs, where residents are encouraged to separate their organic waste from other trash. These programs often provide individuals with composting bins or designate specific areas where organic waste can be dropped off. Additionally, some communities have started composting facilities that process organic waste on a larger scale, making it accessible to those who may not have the means or space to compost at home.

Furthermore, community gardens have emerged as another effective initiative for organic waste recycling. These gardens allow individuals to collectively grow fruits, vegetables, and herbs, while also providing a space to compost organic waste. By utilizing the compost created from their own waste, community gardeners close the loop and create a sustainable system that minimizes waste and maximizes the production of fresh, organic food.

Community initiatives for organic waste recycling not only benefit the environment but also foster a sense of unity and responsibility within the community. By actively participating in these initiatives, individuals become more aware of their environmental impact and develop a stronger connection with their surroundings.

In conclusion, community initiatives for organic waste recycling are crucial for promoting sustainable practices and reducing our ecological footprint. Composting and community gardens are just a few examples of these initiatives that have gained popularity in recent years. As we strive towards a greener future, it is essential for every individual to participate in these initiatives and contribute to the organic waste revolution. Together, we can create a healthier and more sustainable world for future generations.

In recent years, the importance of recycling has become increasingly evident as we grapple with the environmental challenges of our time. One particular area that requires urgent attention is the management of organic waste. Organic waste, such as food scraps and yard trimmings, accounts for a significant portion of the waste generated by households and communities. However, with the right community initiatives, we can tackle this issue head-on and make a positive impact on the environment.

Community initiatives for organic waste recycling play a crucial role in diverting organic waste from landfills and promoting sustainable practices. These initiatives involve the active participation of community members, local organizations, and government bodies. By working together, they create a holistic approach that benefits everyone involved.

One such initiative is the establishment of community composting programs. These programs provide residents with the necessary tools, resources, and education to compost their organic waste at home. Composting not only reduces the amount of waste sent to landfills but also produces nutrient-rich compost that can be used to enrich soil in gardens, parks, and urban green spaces. By actively participating in these programs, individuals can contribute to a more sustainable future while also reaping the benefits of organic gardening.

Another community initiative gaining momentum is the implementation of curbside organic waste collection programs. These programs, often run by local municipalities, provide residents with separate bins for organic waste. The collected waste is then transported to specialized facilities where it is processed into compost. This approach not only simplifies the process for individuals but also ensures that a larger volume of organic waste is diverted from landfills.

In addition to composting programs, community initiatives also focus on raising awareness about the importance of recycling organic waste. Workshops, seminars, and educational campaigns are organized to inform individuals about the environmental impact of organic waste and the benefits of recycling. By targeting schools, community centers, and local events, these initiatives reach a wide audience and encourage them to adopt sustainable practices in their daily lives.

Community initiatives for organic waste recycling are crucial in addressing the environmental challenges we face today. By actively participating in these initiatives, every individual can make a significant difference in reducing waste and promoting sustainability. Together, we can create a cleaner, greener future for ourselves and future generations.

Advocating for Policy Changes and Education

In today's rapidly changing world, it is crucial that we recognize the importance of recycling and take steps to implement effective policies and education programs to tackle the growing issue of organic waste. The need for sustainable solutions has never been greater, and "The Organic Waste Revolution: Composting Solutions for Every Household" aims to provide comprehensive insights into the benefits of recycling and composting for every individual, regardless of their background or lifestyle.

Advocating for policy changes is a key aspect of bringing about a positive impact on our environment. This subchapter highlights the significance of collective action and encourages readers to actively engage with local and national policymakers. By advocating for policies that promote recycling and composting, we can contribute to the creation of a more sustainable future.

One of the main objectives of this subchapter is to emphasize the importance of education. Education plays a crucial role in raising awareness about the benefits of recycling and composting, as well as the detrimental effects of organic waste on our environment. By providing readers with the necessary knowledge and tools, we empower them to become advocates for change within their communities.

This subchapter explores various strategies for promoting education on recycling, including the implementation of educational programs in schools, workplaces, and community centers. It also highlights the importance of public campaigns and awareness initiatives, utilizing social media platforms, and organizing community events to engage a broader audience.

By advocating for policy changes and education, we can foster a culture of sustainability and responsible waste management. The book emphasizes the importance of individual actions and highlights how even small changes in our daily routines can have a significant impact on reducing organic waste.

"The Organic Waste Revolution: Composting Solutions for Every Household" aims to inspire readers to take action and become catalysts for change in their own communities. By advocating for policy changes and education, we can create a world where recycling and composting are not only widely practiced but also celebrated as essential steps towards a healthier environment.

In conclusion, this subchapter serves as a call to action for every individual, emphasizing the importance of advocating for policy changes and education in the context of organic waste management. By working together, we can bring about a revolution in waste management, making a positive impact on our environment and paving the way for a sustainable future.

In this subchapter, we delve into the critical aspects of advocating for policy changes and education to foster an organic waste revolution. It is imperative that every individual understands the importance of recycling and takes necessary steps to contribute to a sustainable future.

Recycling is not just a buzzword; it is a fundamental aspect of preserving our environment. By recycling organic waste, we can significantly reduce the strain on landfills, decrease greenhouse gas emissions, and conserve valuable resources. However, the first step towards achieving these goals is to advocate for policy changes that support and encourage recycling initiatives.

Policy changes play a pivotal role in shaping our waste management systems. Governments must prioritize the implementation of effective recycling policies, provide incentives for composting, and promote education at all levels. By advocating for these policies, we can ensure that organic waste recycling becomes an integral part of our daily lives.

Education is another crucial aspect of fostering a recycling culture. It is essential to empower individuals with the knowledge and tools needed to make informed choices. Educational campaigns can be conducted at schools, workplaces, and community centers to raise awareness about the benefits of recycling organic waste. By educating everyone, regardless of age or background, we can cultivate a sense of responsibility towards our environment and encourage active participation.

Moreover, educational programs should also emphasize the importance of proper sorting and composting techniques. Many individuals are unaware of the correct methods for segregating organic waste from other types of garbage. By educating people about the right practices, we can ensure that organic waste reaches the appropriate composting facilities, thereby maximizing its potential for reuse.

Furthermore, it is crucial to collaborate with local organizations, businesses, and community leaders to drive policy changes and education initiatives forward. By creating partnerships and engaging stakeholders from various sectors, we can establish a united front in advocating for recycling policies and raising awareness.

To truly bring about an organic waste revolution, we must recognize that every individual has a role to play. Whether you are a student, a homemaker, a business owner, or a public servant, your actions matter. By advocating for policy changes and actively participating in educational initiatives, we can collectively create a sustainable future for generations to come.

In conclusion, advocating for policy changes and education is a vital step towards achieving an organic waste revolution. By understanding the importance of recycling, advocating for recycling policies, and actively participating in educational programs, we can pave the way for a greener and healthier planet. Let us come together as a community and make a difference.

Chapter 8: Frequently Asked Questions about Composting

Can I Compost Meat and Dairy Products?

One of the most common questions that arise when it comes to composting is whether or not it is possible to compost meat and dairy products. In this subchapter, we will explore the answer to this question and provide you with valuable insights on how to deal with these organic waste materials in your composting efforts.

Composting is a fantastic way to reduce waste and create nutrient-rich soil for your garden. However, it is important to understand that not all types of organic waste are suitable for composting. Meat and dairy products, in particular, present unique challenges due to their composition and potential for attracting pests.

While it is technically possible to compost meat and dairy products, it is generally recommended to avoid adding them to your home compost pile. These items are rich in fats and proteins, which take longer to break down and can cause unpleasant odors. Moreover, they can attract unwanted pests such as rats and raccoons, which can disrupt the composting process and create a nuisance in your backyard.

Instead of adding meat and dairy products to your compost, it is advisable to seek alternative methods for their disposal. Many municipalities have food waste collection programs that can handle these items separately, ensuring they are properly processed and recycled. If such programs are not available in your area, you can consider using a specialized composting system that can handle these materials effectively and safely.

It is important to remember that the success of composting relies on maintaining a proper balance of materials. To ensure a healthy and efficient compost pile, it is recommended to focus on adding a diverse range of organic waste, such as fruit and vegetable scraps, yard trimmings, coffee grounds, and eggshells. By avoiding meat and dairy products, you can prevent potential issues and maintain a well-balanced composting environment.

In conclusion, while composting is a great way to recycle organic waste, it is best to refrain from composting meat and dairy products in your home compost pile. These items can cause unpleasant odors, attract pests, and disrupt the composting process. Instead, explore alternative disposal methods such as local food waste collection programs or specialized composting systems. By making informed decisions about what to compost, you can contribute to the organic waste revolution and create a sustainable future for our planet.

In our quest to lead more sustainable and eco-friendly lives, composting has emerged as a crucial solution for managing organic waste. However, when it comes to composting, many people are left wondering whether they can include meat and dairy products in their compost piles. In this subchapter, we will explore this question and shed light on the best practices for composting meat and dairy products.

Composting is a natural process that breaks down organic matter into nutrient-rich soil amendment. It involves the decomposition of various materials, such as fruit and vegetable scraps, yard waste, and even paper products. However, when it comes to meat and dairy products, the situation becomes a bit more complicated.

While it is technically possible to compost meat and dairy products, it is generally not recommended for home composting systems. These materials can attract pests, such as rodents and raccoons, which can create a variety of problems. Additionally, the decomposition of meat and dairy products can produce unpleasant odors and attract unwanted insects.

On a larger scale, industrial composting facilities can handle meat and dairy products effectively. These facilities have the necessary equipment and processes to manage organic waste at a larger volume, including the ability to control pests and odor. If you have access to such facilities in your area, consider donating your meat and dairy waste to them instead of including them in your home compost.

However, if you are determined to compost meat and dairy products at home, there are a few things you can do to mitigate the risks. First, make sure to bury these materials deep within your compost pile to deter pests. Second, consider adding a layer of carbon-rich materials, such as dried leaves or wood chips, on top of the meat and dairy waste to help contain odors. Lastly, monitor your compost pile regularly and address any issues promptly to prevent pests and odor problems from escalating.

In conclusion, while it is generally not recommended to compost meat and dairy products in home composting systems, there are options available for those with access to industrial composting facilities.

Remember, the key to successful composting lies in managing a balanced mix of organic materials while avoiding materials that can cause problems. By composting responsibly, we can contribute to the organic waste revolution and create a greener future for everyone.

How Long Does Composting Take?

Composting is a natural process that transforms organic waste into nutrient-rich soil. It is a simple and effective way to reduce the amount of waste that ends up in landfills and to create a valuable resource for our gardens. However, one question that often arises is, "How long does composting take?"

The answer to this question depends on various factors, including the type of materials being composted, the size of the compost pile, and the environmental conditions. In general, composting can take anywhere from a few weeks to several months.

When it comes to the type of materials being composted, some organic waste decomposes faster than others. For example, fruit and vegetable scraps, coffee grounds, and grass clippings break down relatively quickly. On the other hand, materials such as wood chips, branches, and leaves take longer to decompose. By ensuring a good balance between quick-decomposing materials (known as green waste) and slower-decomposing materials (known as brown waste), you can speed up the composting process.

The size of the compost pile also plays a role in determining how long composting takes. A larger pile generates more heat, which accelerates decomposition. However, it is important to note that a pile that is too large may not allow enough oxygen to reach the center, slowing down the process. Therefore, it is recommended to maintain a compost pile that is at least three feet high and three feet wide.

Environmental conditions, such as temperature and moisture, also impact composting time. Microorganisms responsible for breaking down organic matter thrive in warm and moist conditions. Therefore, composting tends to be faster during the summer months compared to winter. It is important to regularly monitor the moisture level of the compost pile and add water if it becomes too dry or cover it with a tarp if it becomes too wet.

In conclusion, the length of time it takes for composting to occur varies depending on the materials being composted, the size of the compost pile, and the environmental conditions. By understanding these factors and making necessary adjustments, you can create a thriving compost pile that transforms organic waste into nutrient-rich soil in a relatively short period of time.

The importance of recycling cannot be overstated. It is essential for preserving the environment, conserving natural resources, and reducing the amount of waste that ends up in landfills. Composting is a crucial part of the recycling process as it diverts organic waste from landfills, where it produces harmful greenhouse gases, and instead turns it into valuable soil amendment.

By composting our organic waste, we not only reduce the strain on our landfills but also create a nutrient-rich soil amendment that can be used in our gardens. This soil amendment, also known as compost, helps improve soil structure, enhances moisture retention, and provides essential nutrients to plants.

Furthermore, composting is a sustainable solution that can be implemented by every household. It does not require any special equipment or complicated processes. All you need is a designated area in your backyard or even a small compost bin for indoor composting.

By composting our organic waste, we can significantly reduce the amount of waste we produce. According to studies, organic waste makes up a large portion of household waste, which can be easily composted instead of being sent to landfills. This reduces the need for additional landfill space and helps combat climate change by reducing the production of methane gas, a potent greenhouse gas emitted by decomposing organic waste in landfills.

In conclusion, composting is a simple and effective way to recycle organic waste and contribute to a more sustainable future. By understanding how long composting takes and the importance of recycling, we can all play a part in the organic waste revolution and make a positive impact on the environment.

Composting is an essential practice in the world of recycling and waste management. It allows us to transform organic waste into nutrient-rich soil, reducing the amount of waste that ends up in landfills. But have you ever wondered how long it takes for this magical process to occur? In this subchapter, we will delve into the timeframe of composting and the factors that influence its duration.

Composting time can vary depending on several factors, including the type of materials used, the size of the pile, the environmental conditions, and the composting method employed. Generally, the process can take anywhere from a few weeks to several months.

When it comes to the materials used, the composition of your compost pile plays a crucial role. Organic waste such as fruit and vegetable scraps, coffee grounds, and yard trimmings decompose faster than harder materials like wood chips or branches. By ensuring a good balance between nitrogen-rich materials (greens) and carbon-rich materials (browns), you can accelerate the decomposition process.

The size of your compost pile is also a determining factor. Smaller piles tend to heat up more quickly, speeding up the breakdown of organic matter. However, larger piles can generate more heat and maintain higher temperatures, which can result in faster composting as well. It's important to find the right balance and adjust the size of your compost pile according to your needs and available space.

Environmental conditions significantly impact the speed of composting. Moisture levels, oxygen availability, and temperature are crucial elements. Adequate moisture allows the microorganisms responsible for decomposition to thrive, while proper aeration ensures their survival. Optimum temperatures between 110°F and 160°F (43°C to 71°C) provide an ideal environment for accelerated composting.

The composting method you choose also affects the timeline. Traditional methods such as hot composting can yield finished compost in as little as three months, while cold composting can take up to a year. Other techniques, like vermicomposting with the help of worms, can speed up the process even further.

In conclusion, the duration of composting depends on various factors, but with proper management and the right conditions, you can obtain nutrient-rich compost in a matter of weeks to months. By understanding the timeline, you can plan your composting activities more effectively and contribute to the organic waste revolution. So, start composting today and join the movement towards a greener and more sustainable future for all.

Can I Compost in Cold or Hot Climates?

Composting is a sustainable and eco-friendly way to manage organic waste, but many people wonder if they can compost in extreme climates. The answer is a resounding yes! Whether you live in a cold or hot climate, composting is possible and can bring numerous benefits to your household and the environment.

In cold climates, composting may slow down due to the lower temperatures. However, with a few adjustments and techniques, you can still successfully compost throughout the year. One way to maintain the composting process in colder temperatures is by insulating your compost pile. You can use materials like straw, leaves, or even old blankets to cover the pile and retain heat. This insulation will help keep the microorganisms active and speed up the decomposition process. Additionally, chopping or shredding your organic waste into smaller pieces before adding it to the pile can accelerate decomposition, even in colder climates. Remember to turn your pile regularly to introduce oxygen and promote the breakdown of materials.

On the other hand, hot climates pose their own challenges when it comes to composting. High temperatures can cause the compost pile to dry out quickly, leading to a lack of moisture for the microorganisms to thrive. To combat this, it's crucial to monitor the moisture levels and add water as needed. You can also choose to compost in a shaded area or use a compost bin with a lid to protect the pile from direct sunlight. Adding more nitrogen-rich materials, such as grass clippings or kitchen scraps, can help balance the carbon-to-nitrogen ratio in the pile and prevent it from becoming too dry.

Composting in any climate is essential for several reasons. Firstly, it reduces the amount of organic waste that ends up in landfills, where it releases harmful greenhouse gases. By composting, you contribute to mitigating climate change and reducing your carbon footprint. Secondly, composting creates nutrient-rich soil that can be used to enrich gardens, lawns, and potted plants. It improves soil structure, retains moisture, and reduces the need for chemical fertilizers. Lastly, composting is a simple and effective way to teach future generations about the importance of recycling and responsible waste management.

Whether you're living in a cold or hot climate, composting is a practical solution for every household. With a little knowledge and adjustments, you can successfully compost organic waste and contribute to the organic waste revolution. Start composting today and join the growing community of environmentally-conscious individuals who are making a positive impact on our planet.

Composting is a simple and effective way to reduce organic waste and contribute to a sustainable environment. However, many people wonder if composting is feasible in extreme climates, such as cold or hot regions. The good news is that composting can be done successfully in both types of climates, with a few adjustments and considerations.

In cold climates, where freezing temperatures are common, composting can be more challenging but not impossible. The key is to insulate and protect the compost pile from extreme cold. To achieve this, it is recommended to use an insulated compost bin or build a compost heap with straw or leaves as a cover. This will help retain heat and prevent the compost from freezing. Additionally, turning the compost more frequently during winter can also help generate heat and accelerate the decomposition process.

In hot climates, where temperatures can soar, composting requires careful management to prevent the pile from overheating and drying out. To ensure successful composting, it is important to keep the compost pile moist by watering it regularly. Adding a layer of straw or shredded paper on top of the pile can help retain moisture. It is also advisable to turn the compost more frequently to aerate it and prevent excessive heat buildup. If the compost pile becomes too hot, it can be cooled down by adding more carbon-rich materials, such as dry leaves or shredded cardboard.

Regardless of the climate, it is essential to maintain a balance of green (nitrogen-rich) and brown (carbon-rich) materials in the compost pile. Green materials include kitchen scraps, grass clippings, and fresh plant waste, while brown materials consist of dry leaves, wood chips, and shredded paper. Achieving this balance is crucial for a healthy composting process and to avoid any unpleasant odors.

Composting in both cold and hot climates is not only beneficial for the environment but also for each individual. By composting organic waste, we divert it from landfills, reducing methane emissions and nutrient runoff. Additionally, composting creates nutrient-rich soil, which can be used to enrich gardens, lawns, and potted plants. This ultimately reduces the need for chemical fertilizers and promotes healthier, more sustainable gardening practices.

In conclusion, composting is possible in any climate, including cold and hot regions. With proper management and a few adjustments, you can successfully compost organic waste and contribute to the organic waste revolution. Let's all embrace composting as a simple yet powerful solution for a greener and more sustainable future.

What Should I Do with Compost during Winter?

As we delve into the organic waste revolution and explore composting solutions for every household, it's crucial to address the question of what to do with compost during the winter months. Many people wonder if composting is still feasible or necessary when temperatures drop, and the answer is a resounding yes! Composting is a year-round activity that can greatly benefit both the environment and your garden.

First and foremost, it's important to understand the significance of recycling and the role composting plays in this process. Recycling is not limited to plastics, paper, and glass; it encompasses organic waste as well. By composting, you are diverting valuable organic materials from landfills, which significantly reduces greenhouse gas emissions and helps combat climate change. Composting is a simple yet effective way to contribute to the importance of recycling and make a positive impact on the planet.

Now, let's address the specific concerns about composting during winter. While the composting process may slow down in colder temperatures, it doesn't come to a halt. Microorganisms responsible for decomposition continue to work, albeit at a slower pace. However, to ensure successful composting during winter, a few adjustments are necessary.

Firstly, consider insulating your compost pile or bin to retain heat. This can be achieved by layering straw, leaves, or other organic materials on top. Additionally, covering your compost pile with a tarp or plastic sheet can help trap heat and protect it from excessive moisture or snow. It's crucial to monitor the moisture content regularly and avoid overwatering, as excessive moisture can hinder the decomposition process.

If you live in an area with extremely cold temperatures, you may opt for an indoor composting system, such as a worm bin. These bins are compact, odorless, and perfect for apartment dwellers or those with limited outdoor space. They can be kept indoors, providing a controlled environment for composting throughout the year.

Remember, the key to successful composting during winter is patience. It may take longer for your compost to fully decompose, but the end result will be worth the wait. Come spring, you'll have nutrient-rich compost ready to nourish your garden and help your plants thrive.

In conclusion, composting during winter is not only possible but also crucial for the organic waste revolution and the importance of recycling. By making a few adjustments and being patient, you can continue composting throughout the year, reducing waste, and contributing to a healthier planet. So, embrace the cold season and let your composting efforts shine even when the temperatures drop.

During the winter months, many people wonder what to do with their compost. The cold weather and freezing temperatures can make composting seem challenging, but fear not – there are still plenty of options to keep your composting efforts going strong even in the midst of winter.

One option is to insulate your compost pile. By covering it with a layer of straw, leaves, or even a tarp, you can help retain heat and protect it from freezing. This will ensure that the decomposition process continues, albeit at a slower pace. Remember to remove the cover periodically to allow for aeration and moisture control.

If maintaining an outdoor compost pile seems too challenging during winter, consider starting an indoor composting system. There are many compact and odor-free options available, such as worm composting or vermicomposting. These systems can easily be set up in a small corner of your kitchen or garage. Not only will they continue to convert your organic waste into nutrient-rich compost, but they also provide a fascinating educational experience for children and adults alike.

Another option for composting during winter is to find a local community garden or composting facility that accepts organic waste. Many of these facilities have large compost piles that are managed to maintain optimal temperatures even in cold weather. By donating your compost to these organizations, you can ensure that your organic waste continues to be recycled and transformed into valuable soil amendments.

If you have a garden, you can also use your compost to protect your plants during winter. Spread a layer of compost around the base of your plants to provide insulation and protect them from the harsh cold. As the compost breaks down slowly over time, it will also release nutrients that will benefit your plants come springtime.

Remember, regardless of the season, composting is an essential practice for every household. It not only reduces the amount of waste sent to landfills but also enriches the soil and supports a healthy ecosystem. So, don't let the winter deter you – find the best option that suits your needs and continue composting throughout the year.

In conclusion, composting during winter can be done in various ways, from insulating your outdoor pile to starting an indoor composting system or donating to community gardens. By finding the best approach for your situation, you can ensure that your organic waste is recycled and contribute to the overall importance of recycling in our society. Keep the organic waste revolution alive, even in the coldest months!

Chapter 9: Inspiring Success Stories in Organic Waste Recycling

Household Success Stories

In this subchapter, we will explore some inspiring household success stories that demonstrate the power and potential of composting solutions in our everyday lives. These stories highlight the importance of recycling organic waste and how it can positively impact both our environment and our communities. By sharing these success stories, we hope to inspire and motivate every individual to join the organic waste revolution.

1. The Smith Family: The Smiths, a family of four, decided to embark on a composting journey after realizing the amount of food waste they were generating. They set up a small compost bin in their backyard and started composting their kitchen scraps and yard waste. Within a few months, they saw a significant reduction in the amount of waste they were sending to the landfill. The Smiths were amazed at how their compost helped nourish their garden, resulting in a bountiful harvest of fresh vegetables. This success story not only reduced their carbon footprint but also inspired their neighbors to start composting.

2. The Green Office: A small startup implemented an office-wide composting program after learning about the benefits of recycling organic waste. By providing compost bins in the break room and educating employees about the importance of composting, they were able to divert a substantial amount of waste from the landfill. The company then used the compost to fertilize the office plants, creating a healthier and greener workspace. This success story not only reduced

their environmental impact but also fostered a sense of community and eco-consciousness among the employees.

3. The Urban Gardener: Sarah, a city dweller with limited space, discovered the power of composting and its potential for urban gardening. She started composting her kitchen scraps using a small indoor compost bin. The nutrient-rich compost she produced helped her cultivate a thriving rooftop garden, growing vegetables and herbs that provided fresh and organic produce for her family. Sarah's success story not only showcased the possibilities of composting in urban settings but also inspired her neighbors to utilize their limited spaces for sustainable gardening.

These household success stories demonstrate that composting solutions are accessible to everyone, regardless of their living situation or lifestyle. By recycling organic waste, we can reduce our environmental impact, enrich our soil, and cultivate a sustainable future. Whether you have a backyard or a small apartment, whether you're an individual or a business, composting is an achievable and rewarding practice that benefits both you and the planet. Join the organic waste revolution and be a part of these inspiring success stories.

In this subchapter, we delve into inspiring stories of households that have successfully embraced the organic waste revolution and adopted composting solutions. These stories highlight the importance of recycling and demonstrate how simple changes in our daily lives can have a meaningful impact on the environment.

Meet the Johnsons, a family of four who decided to take control of their organic waste and transform it into nutrient-rich compost for their garden. Before they started composting, their trash bin was full of food scraps, yard waste, and other organic materials. They realized that by composting, they could drastically reduce their waste going to the landfill while also enriching their soil. Today, their garden flourishes with healthy plants and vibrant flowers, thanks to the compost they produce at home.

Another inspiring success story comes from the Smiths, a couple who lives in an urban apartment. Initially, they were skeptical about composting due to limited space. However, with some research and creativity, they discovered the power of vermiculture - composting with worms. They set up a small worm bin under their kitchen sink and started feeding the worms with their food scraps. To their surprise, the worms efficiently transformed their waste into valuable vermicompost. Not only did this solution reduce their carbon footprint, but it also provided them with a sustainable source of fertilizer for their balcony garden.

The Thompsons, a retired couple, found their passion for composting after attending a local workshop on sustainable living. They were introduced to the concept of a community composting program, where neighbors collect their organic waste in a central location to be composted collectively. The Thompsons became active members of this program, and soon their enthusiasm spread throughout their community. Today, their neighborhood is recognized for its commitment to composting, reducing waste, and building a stronger sense of community.

These stories showcase the importance of recycling and how individuals from all walks of life can contribute to the organic waste revolution. By composting, these households have not only minimized their environmental footprint but also reaped numerous benefits, such as healthier gardens, reduced landfill waste, and a sense of pride in their sustainable practices.

Whether you live in a spacious house with a backyard or a small apartment with limited space, there is a composting solution for you. The organic waste revolution is not just for those with green thumbs, but for everyone who cares about the planet. By taking small steps to recycle our organic waste, we can collectively make a significant impact on the environment and create a more sustainable future for generations to come.

Community and Citywide Success Stories

In this subchapter, we will take a closer look at the incredible success stories that have emerged from communities and cities around the world who have embraced the organic waste revolution and implemented effective composting solutions. These stories highlight the importance of recycling and demonstrate how small changes at the household level can have a significant impact on the environment and the overall well-being of a community.

One remarkable success story comes from the city of San Francisco, which has become a leader in composting and recycling. Through a combination of educational programs and comprehensive waste management systems, San Francisco has achieved an impressive 80% waste diversion rate. This means that 80% of the city's waste is being recycled or composted instead of ending up in landfills. By implementing mandatory composting and providing residents with compost bins, the city has not only reduced its greenhouse gas emissions but has also created a valuable resource in the form of nutrient-rich compost that is used to enrich local soil.

Another inspiring example comes from a small community in rural India, where a group of women started a composting initiative. With limited resources and minimal infrastructure, they managed to create a thriving composting system that not only reduced waste but also provided a much-needed source of income for the community. The compost they produced was sold to local farmers, improving soil fertility and crop yields. This initiative not only helped the environment but also empowered the women involved and strengthened the sense of community.

These success stories demonstrate that recycling and composting can bring about positive change on both a local and citywide scale. They show that even small actions, such as separating organic waste from other garbage or starting a small compost pile in your backyard, can make a difference. By recycling and composting, we can reduce the amount of waste that ends up in landfills, conserve natural resources, and decrease greenhouse gas emissions.

Furthermore, these success stories highlight the importance of education and community involvement. They demonstrate that raising awareness about the benefits of recycling and providing the necessary tools and infrastructure can lead to substantial changes in waste management practices. By sharing these success stories and celebrating the achievements of communities and cities, we hope to inspire others to take action and become part of the organic waste revolution.

In conclusion, the success stories of communities and cities that have embraced composting and recycling provide tangible evidence of the importance of these practices. They demonstrate that every household has the power to make a positive impact on the environment and contribute to a more sustainable future. By learning from these stories and implementing composting solutions in our own lives, we can join the growing movement towards a greener, healthier planet.

Title: Community and Citywide Success Stories: Inspiring Tales of Recycling and Composting

Introduction:
In this subchapter, we delve into the amazing success stories from

various communities and cities around the world that have embraced composting and recycling as a way to tackle the organic waste crisis. These stories demonstrate the power of collective action and inspire individuals and communities to take charge of their organic waste management. By sharing these success stories, we hope to highlight the importance of recycling and motivate every reader to contribute towards a sustainable future.

1.   The   Green   City   Project: Let's begin by exploring the transformation of a once-polluted city into an eco-friendly haven. The Green City Project, initiated by community leaders in collaboration with local authorities, successfully implemented a comprehensive organic waste management system. Through community engagement, education, and infrastructure development, the project achieved a remarkable 80% reduction in organic waste sent to landfills within just two years.

2.   The   Neighborhood   Composting   Initiative: In a small suburban neighborhood, a group of enthusiastic residents took matters into their own hands. They initiated a neighborhood-wide composting initiative, providing compost bins to each household. Through regular workshops and educational campaigns, they empowered residents to compost their organic waste. This grassroots movement not only reduced landfill waste but also created a strong sense of community and environmental stewardship.

3.   The   School   Recycling   Program: Schools are powerful agents of change. The School Recycling Program in a major city mobilized students, teachers, and parents to recycle and compost organic waste generated within school premises. With the

installation of recycling bins and the incorporation of environmental education into the curriculum, students developed a deep understanding of the importance of recycling and took these practices back to their homes, creating a ripple effect throughout the community.

4.   The   Citywide   Composting   Challenge: In an effort to become a zero-waste city, a metropolitan area launched a citywide composting challenge. The initiative encouraged businesses, residents, and institutions to compete in diverting organic waste from landfills. The challenge offered incentives, such as tax breaks and recognition, to participants who achieved the highest reduction rates. This friendly competition resulted in a significant reduction of organic waste and inspired other cities to adopt similar initiatives.

Conclusion:

These stories highlight the transformative impact of recycling and composting initiatives on communities and cities. They prove that change starts at the grassroots level and can grow into a citywide movement. By embracing these success stories, we can all play a role in creating a sustainable future. Whether you are an individual, a community leader, or a city official, there are endless opportunities to make a difference. Let these success stories inspire you to take action and join the organic waste revolution. Together, we can build a greener, cleaner future for all.

Innovations in Organic Waste Recycling

Recycling has become an increasingly important aspect of our lives, and the same holds true for organic waste. As we strive to protect our environment and reduce our carbon footprint, finding innovative solutions for recycling organic waste has become crucial. In this subchapter, we will explore the various advancements in organic waste recycling and how they contribute to the larger goal of sustainability.

One of the most notable innovations in organic waste recycling is the development of home composting systems. These systems allow every household to take an active role in recycling their organic waste. With the rising popularity of urban gardening and sustainable living practices, home composting has gained significant attention. By recycling kitchen scraps, yard waste, and other organic materials, individuals can create nutrient-rich compost that improves soil health and reduces the need for chemical fertilizers.

Another exciting innovation in organic waste recycling is the use of anaerobic digestion. Anaerobic digesters are large-scale systems that break down organic waste in the absence of oxygen, producing biogas, a renewable source of energy. This technology has revolutionized waste management in many industries, including agriculture, food processing, and wastewater treatment. By harnessing the potential of anaerobic digestion, we can convert organic waste into valuable resources while reducing greenhouse gas emissions.

Advancements in technology have also led to the development of automated sorting systems for organic waste. Traditional recycling processes often struggle to efficiently separate organic materials from other waste streams. However, with the introduction of cutting-edge technologies like optical sorting and artificial intelligence, we can now sort organic waste more accurately and effectively. These systems not only streamline the recycling process but also help maximize the potential for resource recovery.

Moreover, innovations in organic waste recycling have extended beyond traditional composting methods. Researchers and entrepreneurs are exploring alternative uses for organic waste, such as the production of bioplastics, biofuels, and bio-based chemicals. These innovative approaches not only divert organic waste from landfills but also contribute to the creation of a circular economy, where waste is transformed into valuable resources.

In conclusion, innovations in organic waste recycling play a vital role in addressing the importance of recycling for everyone. From home composting systems to anaerobic digestion and advanced sorting technologies, these innovations offer practical solutions for managing organic waste effectively. By embracing these innovations, we can reduce our environmental impact, conserve resources, and create a more sustainable future for all.

Chapter 10: Conclusion and Call to Action

In this final chapter of "The Organic Waste Revolution: Composting Solutions for Every Household," we come to a crucial point where we summarize the key takeaways and issue a call to action to every individual. Throughout this book, we have explored the importance of recycling organic waste and the immense benefits it holds for our environment, our health, and our future generations.

Recycling is not just a responsibility of a few; it is a duty that falls upon each and every one of us. The impact of our actions, or lack thereof, can either contribute to the degradation of our planet or help build a sustainable future. It is time for us to rise up and take charge of our waste management practices.

We have learned that organic waste, which makes up a significant portion of our daily trash, can be transformed into valuable compost through the process of composting. Composting not only reduces the amount of waste that ends up in landfills but also creates nutrient-rich soil amendments that can be used to grow healthier plants and vegetables.

By diverting organic waste from landfills, we can significantly reduce greenhouse gas emissions. Landfills are notorious for producing methane, a potent greenhouse gas that contributes to climate change. Composting, on the other hand, promotes the growth of beneficial bacteria that break down organic matter aerobically, minimizing methane production.

Furthermore, recycling organic waste reduces the need for chemical fertilizers, which can have detrimental effects on soil health and water quality. By embracing composting, we can create a circular system where our waste becomes a valuable resource, closing the loop on our consumption patterns.

The call to action is simple: start composting today! Regardless of whether you live in an apartment or a house, there are composting solutions available for every household. From small-scale indoor composters to outdoor bins, there is an option that fits your lifestyle and space constraints.

Educate yourself and your community about the benefits of recycling organic waste. Spread the word about composting, share your experiences, and inspire others to join the movement. Together, we can make a significant difference in reducing waste, conserving resources, and protecting our planet.

Remember, the importance of recycling goes beyond just our immediate surroundings. It is about ensuring a healthier and more sustainable future for generations to come. Let us embrace the organic waste revolution and be the change we wish to see in the world.